God, If You're Not Up There, I'm F*cked

TALES OF STAND-UP,

SATURDAY NIGHT LIVE,

AND OTHER

MIND-ALTERING

MAYHEM

God, If You're Not Up There, I'm F*cked

Darrell Hammond

HARPER LUXE

An Imprint of HarperCollinsPublishers

To the boys from Hell's Kitchen:
Marty Hennessy, Bobby Spillane,
and Big Mike Canosa

And to Myrtise

HarperCollins books may be purchased for educational, business, or sales promotional use. For information please write: Special Markets Department, HarperCollins Publishers, 10 East 53rd Street, New York, NY 10022.

FIRST HARPERLUXE EDITION

HarperLuxe™ is a trademark of HarperCollins Publishers

Library of Congress Cataloging-in-Publication Data is available upon request.

ISBN: 978-0-06-208875-8

11 12 13 14 ID/OPM 10 9 8 7 6 5 4 3 2

Contents

Author's Note

What follows in these pages is my life as best I remember it. Some things happened a long time ago, some of them happened when I was under the influence of alcohol, drugs, and/or any number of hard-core psychopharmaceuticals prescribed for me by about two dozen shrinks, so I'm not always perfect with the details. I did the best I could. Some names and details have been changed to protect people's privacy; while I was proud to know them, I can't be sure the feeling was mutual.

Prologue

You know what's worse than being in rehab? Being in rehab over the holidays. You know what's worse than that? Being in a rehab that doesn't allow smoking. I mean, what the fuck? Addicts smoke. If we can't drink, we can't shoot up, and we can't ride the lightning bolt, at least we can smoke.

I was sent to the Sanctuary, a few miles north of New York City, via ambulance in the fall of 2010 after getting drunk and trying to cut my arm off with a large kitchen knife. It is one of the best psychiatric and rehabilitation facilities in the country. I was put in the "celebrity ward," which drew its share of boldfaced names—award-winning actors, sports stars, European royalty—but there are also wards for specific mental illnesses—depression, schizophrenia, eating disorders—and a criminal unit filled with dealers, streetwalkers, thieves, and assorted other miscreants who were there by order of the court.

Deprived of my freedom, separated from my family, I was one of the lucky ones being given yet another chance. It sucked.

I've been hospitalized or shipped off to rehab so many times that I've honestly lost count. But each one had its own particular brand of hell. The program at the Sanctuary proudly boasted of its success in bringing addicts back to health while generously providing all the butter-laden cookies and cream-filled pastries we could cram into our alcohol-starved, sugar-craving mouths. Hell, I put on twelve pounds in the first three weeks trying to get "healthy."

Meanwhile, the ferret-faced floor wardens were always looking to bust us for any infraction. There was one nurse there, an attractive, muscular woman in her forties who we called Strap-On because she was constantly reaming someone for some petty crime. She and one of the "tough love" counselors busted us for smoking numerous times. Each room had its own bathroom, and when she caught me hanging out the window of mine with a lit cigarette, she announced it loudly to all within earshot, "He's smoking in the bathroom!" as though she'd discovered Satan carving his initials in a church pew.

So to avoid her wrath, and if it wasn't too cold or snowing, the smokers would wander out of the building, down a flagstone path that wound across the finely manicured grounds, to The Tree, the worst-kept secret in the place. An ancient cedar encircled by a layer of dead butts like some weird white-and-tan mulch, it was wide enough and tall enough and just far away enough to hide a grown man getting his nicotine on.

By Thanksgiving Day, I'd been in this rehab for three weeks, and I'd run out of cigarettes. My family wasn't speaking to me, and my friends were all doing their own thing for the holiday, so no pumpkin pie and stuffing for me. Some of the other patients were spending time with their loved ones in the glassed-in sunroom, awkwardly trying to act "normal." I figured I'd take a stroll by The Tree to see if any of the other smoker derelicts were there, so I tiptoed past and out the door.

Bingo. Annabelle, a stunning mocha-skinned hooker from Philly, was sucking on a Marlboro Red. Annabelle's lawyer had convinced a judge when she got done for possession that she needed a doctor more than a jail, so she wound up here instead of the Bedford Hills Correctional Facility for Women, where such lovelies as Amy Fisher, the Long Island Lolita who shot her lover Joey Buttafuoco's wife in the face, have done hard time. Annabelle had been at the Sanctuary about a week.

"Hey, Joe." Everybody knew me as Joe. I hadn't been on TV in a while—my last appearance on *Saturday Night Live*, a cameo as Arnold Schwarzenegger on "Weekend Update," was a year earlier—but I wasn't in the mood to be recognized while I got myself sorted out, so I checked in under a false name. Unlike certain celebrities who like to share their meltdowns with Matt Lauer or TMZ, I prefer to bring the heavens crashing down around me in private.

"Hi there," I said. Gorgeous as she was, I couldn't take my eyes off her cigarette.

Annabelle caught me looking. "You want a drag?"

I took the butt from her extended hand. The cherry red lipstick on the filter was definitely not my color, but I didn't care. Those two puffs were about the best I ever had.

"Thank you," I said. "I owe you." I noticed her hand was trembling when she took the cigarette back from me. From the cold or withdrawal, I couldn't tell which.

"No problem, Joe," she said. Then, smiling, "Now say it like Bill Clinton."

CHAPTER ONE
The Hall

Studio 8H, 30 Rockefeller Center
New York City
1995

To say it's intimidating to walk into 30 Rockefeller Center to audition for *Saturday Night Live* is one of the century's greatest understatements. The building itself, once known as the RCA Building until GE bought the company and NBC along with it, is one of the city's great landmarks, built during the Depression in classic Art Deco style. You could get dizzy looking up at the Josep Maria Sert mural *Time* on the ceiling above the main entrance. Thank God it was summer, because if the enormous Christmas tree had been up out front, I'd probably have passed out.

Trying to ignore the hordes of tourists lined up to take the NBC tour, I checked in at the security desk—*Yes, Mr. Hammond, here's your pass, go on up, they're expecting you*—and stepped into the same elevator that for two decades had ferried a seemingly endless cavalcade of comedians to stardom.

I got out on the eighth floor and was escorted to makeup, where a lovely young lady dabbed me with powder to douse the shine of nervous sweat on my forehead. At least I had a few months of sobriety under my belt, so I didn't have withdrawal shakes. Although I could have killed for a slug of gin right about then.

When I'd been sufficiently fluffed and primped, I was led into the theater that I'd fantasized about forever, Studio 8H, or the Hall, as I call it, where legends like George Carlin, Buck Henry, and Andy Kaufman had performed, a few feet from where the Rolling Stones and David Bowie have played, and where Lorne Michaels, who hatched this comedy phenomenon a generation earlier to replace weekend reruns of *The Tonight Show*, was sitting on a chair in front of me.

I almost said, "You know what? I'm thirty-nine years old. I'm on lithium. Do you know what lithium is for? If I may quote the National Library of Medicine at the National Institutes of Health:

Lithium is used to treat and prevent episodes of mania (frenzied, abnormally excited mood) in people with bipolar disorder (manic-depressive disorder; a disease that causes episodes of depression, episodes of mania, and other abnormal moods). Lithium is in a class of medications called antimanic agents. It works by decreasing abnormal activity in the brain.

"So yeah, it's too late, and I'm too fucking scared, and apparently I have abnormal activity in my brain. Thank you. Good-bye."

Lorne looked at me and said, "Are you okay?"

"I think so," I lied.

Then he smiled at me. *Fuck it, now I have to go through with it.* I had been asked to do ten minutes, which is an eternity. I proceeded to peel off every impression, like Phil Donahue speaking Spanish, that I could pull together in the short amount of time that I'd been given to prepare. But really, I'd been preparing for the previous twelve years for a moment like this.

When I left, I thought, If my life ends right now, it's okay. I was in that theater. Lorne Michaels was there. And I performed well, despite my terror. Whether I was any good or not was immaterial. I knew I wasn't going to get called back.

And yet I was. Lorne wanted to see if I had any more impressions, so I came up with Ted Koppel in German and another ten or so and did it all over again a week later.

I guess I did okay, because then Lorne wanted to see me move on my feet. He took me to the Comic Strip. Fuck, *that* place? *Really?* I'd been dismissed by a lot of New York clubs, but the Comic Strip held a special place in my pantheon of rejection.

When I auditioned there in 1990, I had as great a set as I'd had up to that point. I totally killed. Afterward, Lucien Hold, the manager of the club, who was renowned for having discovered comic greats like Jerry Seinfeld, Chris Rock, and Adam Sandler, sat me down in a booth up front.

"How old are you?" he asked.

I told him I was thirty-four. He didn't flinch.

He pointed to pictures on the wall of comedians who had worked there. "Look at these faces. They're stars. That's what we're about here. Stars."

Lucien was smiling at me.

"They have *it*."

Could my luck be changing?

"And I don't think you have *it*," he said.

Apparently my luck wasn't changing in any way whatsoever.

"I don't see any reason why you should come back here or call here again." He stood up and walked away without so much as a "Good-bye, thanks for coming."

I went home with absolutely no reason to believe that I was ever going to make it. I had only enough money for a subway token. It was one of those horribly cold February New York nights, and I took the train back to my hovel in Brooklyn. I even slipped on the ice on the sidewalk outside my apartment. It was perfect, going home without hope. I sat in the dark, smoking cigarettes. If I'd had any money, I'd have gotten drunk.

It was kismet that, five years later, Lorne would have me go there for part of my audition, unwittingly giving me a dose of cosmic payback. With Lorne watching and Lucien Hold hovering nearby, I got onstage and, once again, I killed, although with so many performances and impressions since then, I no longer remember exactly what I did. As much as I would have liked to tell Lucien what I thought of him, I figured that performance for Lorne was as much of a fuck-you as I needed.

The next challenge was dinner with Lorne and his producer, Marci Klein, Calvin Klein's daughter, who to this day works as a producer on *Saturday Night Live* as well as with Tina Fey on *30 Rock*. Marci had chosen a restaurant over on the West Side near Broadway. It was a casual evening, and we just swapped stories. The problem is, the highlight of one of my stories might be, "And then when we got to the store, they didn't have any long-handled spoons!" and Lorne's would be something like, "When I was sitting on the Berlin Wall with Paul McCartney . . ." I was never going to be able to compete with his material, but somehow I made it through the evening without humiliating myself.

A few weeks later, I was lying on my futon on the uneven floor of my apartment in Hell's Kitchen, where the night before a small furry creature had run up the side of my head, stomped on my face, and then run back down the other side. My wife was with me when the phone rang. I don't know why, but we looked at each in that meaningful way they do in the movies. It was my manager, Barry Katz, and my agent, Ruth Ann Secunda, calling at the same time, which never happened.

I got the job.

Holy shit.

My wife and I decided to celebrate by opening a bottle of champagne and dancing in the fountain in front of the Plaza Hotel.

Okay, no, we didn't. What we really did was run like crazy from my apartment on Forty-eighth Street and Tenth Avenue down to Forty-fifth Street between Eighth Avenue and Broadway to the Imperial Theatre, where *Les Miz* was playing. I loved that show. I'd seen it about ten times. I couldn't get enough of it; we bought the most expensive tickets they had left. I reckoned that play is my life story—unjustly treated by life, resolutely angry, but things kind of work out, and along the way there's a little bit of love and light and, not for nothing, a couple of bucks in it too. That's my Tenth Avenue synopsis of one of the great literary works of all time.

A few days later, I was having dinner at Umberto's Clam House down in Little Italy, and I ran into Colin Quinn, whom I'd met years earlier when I'd been hired, then fired, as his warm-up guy on the MTV game show *Remote Control,* which he hosted in the late 1980s.

"Hey, what's up?"

"I just got *Saturday Night Live!*"

"Me too!"

On Monday, September 25, 1995, I reported to work for the twenty-first season of *Saturday Night Live.*

As I walked through those halls and saw those photos on the wall of the greats who had worked there—the late John Belushi, Gilda Radner, and Chris Farley; Dan Aykroyd, Eddie Murphy, Dana Carvey, Mike Myers, Molly Shannon—I couldn't wrap my lithium-quenched mind around the fact that this was really happening. Two decades of comic genius *and me?* It was tremendous validation, and yet I was certain it was a cosmic joke of some kind. Maybe the antidepressants were making me hallucinate?

I was assigned an office on the seventeenth floor with, who else, Colin Quinn, who had been hired that year as a writer and featured player. We hung out there until an intern knocked on the door and yelled, "Pitch!"

The pitch meeting in Lorne's office down the hall is a little bit tradition and a little bit meet-and-greet, where whichever legend of stage or screen or music or sports or politics is hosting that week is welcomed by the writers and the cast. Lorne sits behind his desk, the host sits in a chair by the desk, and everyone else sits wherever they can squeeze in, including the floor. Lorne's office was plenty big, but it's a shitload of people who crammed in there.

Lorne's right-hand men in these festivities, alongside Marci Klein, were *SNL* producers Mike Shoemaker and Steve Higgins. These guys had, and still have, to be able to do all the jobs on the show—like a restaurant manager who can cook, wait tables, and make a nice Caesar dressing. They had to know how to write a joke, manage people, craft a sketch, and above all they had to be fucking funny. Higgins is great with impressions and helped me build almost every impression I would do on the show.

Mariel Hemingway was the host of my first show. I remember thinking at the time, *She's had a conversation with Woody Allen, hell, she's kissed Woody Allen, and I'm sitting just a few feet from her. And she's more beautiful than anyone has ever been in the history of people.*

Do my socks match?

Fuck.

Even major stars are often a little intimidated when they walk into those offices, but Lorne makes sure the *SNL* crew around them is very hands-on in the most unintrusive way possible. Everybody is extremely welcoming, and any thought the host might have, the slightest grievance, the slightest knitted brow, is addressed clearly and immediately. And the host has tremendous say in what the show will be that week.

During the pitch meeting, everybody throws out ideas to the host about what they might like to do for that week's show. If you don't have an idea, it's entirely okay to make up something, even if it's hideous. The staff laughs, but often the poor host sits there thinking, What? Grecian Formula 44 on toast? What?

I tossed out a crazy idea for a cold open—that's the sketch at the start of the show that always ends with, "Live from New York, it's *Saturday Night!*" The show had been receiving a lot of criticism for not being as funny in recent seasons as the nation thought it should be, and Lorne was being battered a little bit in the press for the way the show had gone downhill. So I had this crazy idea of doing a *Wizard of Oz* sketch in which the bad press was just a dream. It was completely ridiculous, although everyone was very kind about it.

It was the first of nearly three hundred weeks I would spend at *SNL*, and lithium isn't exactly *Ginkgo biloba* when it comes to memory, so I'm going to admit the details of that first week are a bit murky all these years later, but here goes.

After the pitch meeting the cast and writers, per custom, spent some time with Mariel, chatting, swapping compliments, and drinking coffee. Some writers started to work on the sketches that got the nod during the pitch meeting, and the rest of us went home.

On Tuesday, people started coming in around noon. Tradition dictates that the host visit with the writers in their offices to talk about proposed sketches. A lot of the writers are Emmy winners, and it's really an honor for the host as much as anything. Meanwhile, the writing began in earnest, so a lot of people would end up staying all night working, including Lorne. The cast members conferred with the writers, and each one participated in putting together anywhere from five to ten sketches.

My role on the show was a little different from the rest of the cast's. I wasn't very good at coming up with sketch ideas, but that's not why I was there. I was a field-goal kicker. You need a voice? I was the guy who could kick that football. I didn't know how to punt, pass, or tackle, but I could kick. So I came in on Wednesday mornings around 10:00 a.m. Sometimes I'd have gotten a call Tuesday night that would be a tip-off: "Can you do this guy?" But lots of times they would simply assign a role to me, and I would walk in having never heard the voice. That first week they gave me Ted Koppel in a *Nightline* sketch in which Koppel interviews Republican presidential candidates Colin Powell (Tim Meadows) and Bob Dole (Norm Macdonald). I usually had four or five hours to study videotapes I got from the research department and cobble together some semblance of a voice before read-through with all the cast and crew, which happened Wednesday afternoon.

You know how they open the gate at a rodeo and the bull comes out seething with energy and fury? That's the kind of mindset you have to have at read-through, because that's how hard it was. But it was all part of it, and as a performer you wouldn't have it any other way. A tennis player expects Wimbledon to be tough, and it is. Anywhere from thirty-five to fifty sketches were presented, which is three or four times what we'd end up with. It took a few hours to get through them all.

After read-through, Mariel complimented me on my Koppel impression. I thought, This is all pixie dust.

Right after the meeting, Lorne and the head writer and the producers met with Mariel to make the first cut, based chiefly on how many people laughed at each sketch during read-through. Lorne and the host had the last word on what stayed in and what got tossed.

Between read-through and picks, you might find yourself loitering around with hosts like: Robert De Niro, Senator John McCain, Ben Affleck, Jennifer Aniston, Sir Ian McKellen, Snoop Dogg, and Derek Jeter. For starters.

Later that night, an intern came by all the offices and yelled, "Picks!" That's when we found out which ten or twelve sketches had survived. It was always a drastic cut that day, and, per usual, a few people took a hit to the solar plexus, but there was no time to nurse hurt feelings. Costume and makeup started getting designed Wednesday night after picks. On Thursday, the writers revised or reworked the sketches that needed it. "Weekend Update" started to get pulled together based on what was in the news that week. On Thursday and Friday we blocked the show—that's figuring out the physical part of the performance, where everybody stands and how they move—while the writers oversaw the costuming and set design of their sketches. The writers also worked on Mariel's monologue.

Saturday afternoon we did a run-through for Lorne, during which the writers made notes for further changes—to dialogue, costume, blocking, and set design. The cast was usually in at least partial costume—wigs and costume, if not full makeup. In the two or three hours between the end of the run-through and the dress rehearsal at 8:00 p.m. (which is done in front of a live studio audience, although not the same one that will be seated for the live show at 11:30), the sketches were tweaked yet again, and new scripts distributed; each version of a script was a different color so everyone knew which was the most current. Sometimes the order of sketches was changed; often, sketches were cut. I went back to my dressing room to refine my Koppel impression and grab some dinner and a nap.

For dress, the cast got into full costume and makeup. The writers paid a lot of attention to how the audience reacted. The show was running at two hours, so when dress was over at ten o'clock, everybody headed up to Lorne's office on the ninth floor and waited for "Meeting!" to be called. This was when Lorne and the writers made the final cuts to get the material down to ninety minutes, as well further tweaks to the surviving sketches.

When the meeting ended, it was nearly eleven o'clock, so the cast hustled back downstairs to get back into hair and makeup (to make sure we didn't ruin the costumes or the handmade hairpieces, we would take them off between performances throughout the day). The writers got to work on making the changes. A posse of interns hits the bank of copy machines to churn out the revised scripts.

It's incredibly confusing, and the frenzy continues all the way until 1:00 a.m. If the show is running long, further sketches might be cut while we're on air. And adding to the stress, for me at least, was the constant presence of seriously famous people who came by to say hello to Lorne and watch the goings-on from a discreet spot on the floor. Over the years, a who's who of boldfaced names stopped by, but the audience can't see that Paul McCartney is standing right in front of you, watching you do your sketch, or that Yankees All-Star A-Rod is there with Kate Hudson, looking at you with an expression that says, *Be spectacular. Be like us.* The audience also doesn't know that you may not have seen the script before now, that it might have been rewritten on the way downstairs from the ninth-floor meeting to the eighth-floor studio.

Not everyone can take it, but Lorne picks people who can operate in this biosphere. He doesn't just hire talented people, he hires fast people. Some really fabulous players were on for only one year, and some fabulous people didn't make it even that far. The late Bernie Brillstein, who was my manager as well as Lorne's, once told me that Lorne has the greatest mind in comedy. And the truth is, at the heart of the show's greatness is a mystery that really only Lorne understands.

Intimidated as I was, I was deliriously happy to be along for the ride.

During Mariel Hemingway's monologue, she wandered backstage to introduce both new and returning cast members. I was one of several new people that year, along with Jim Breuer, Will Ferrell, David Koechner, Cheri Oteri, and Nancy Walls, plus returning players Norm Macdonald, Mark McKinney, Tim Meadows, Molly Shannon, and David Spade. Mariel had recently had a guest-starring role as a lesbian on *The Roseanne Show*, so in this bit she breezed by all the men and kissed all the women full on the mouth, even the show's new director, Beth McCarthy. It was pretty hot.

Right after the monologue, there was a pretaped gag commercial for A.M. Ale that had me sucking down a 40 with my breakfast. How fitting for my first day on national television.

Later in the show I debuted my Ted Koppel, which, regardless of Mariel's lovely words after read-through, didn't get a single laugh on air. I got his voice dead right, but it turned out that was exactly the problem: I did it as a straight impression. Most of the voices I'd learned over the years had been in the context of obvious jokes, but I was intimidated by the seeming seriousness of the *Nightline* sketch. Lorne used to say, "Start accurate, and then exaggerate the impression to make it funny."

By the time we reached the good nights, it felt like I'd been through a year of my life in the last twelve hours. There were hundreds of hurdles and little pitfalls and artistic souls littering the halls like umbrella carcasses along a New York street after a hard rain.

But it didn't matter. This was nirvana. Standing on the stage next to Mariel Hemingway at the end of the show, waving to the studio audience, I thought, This is just as good as playing for the Yankees.

I was so revved up that I went right back in to work Sunday morning, when the offices were empty, and worked on my Koppel. I didn't know if I'd do him again, but I wasn't taking any chances.

The following Saturday, I was in my first cold open. I played sportscaster Bob Costas, although I didn't say "Live from New York, it's *Saturday Night!*" I had the flu and it didn't go as well as I wanted, which he needled me about good-naturedly the first time I met him.

Mariel Hemingway must have had a great time when she hosted, because she came back for a cameo that week (she served Lorne coffee). When I took the stage with my fellow cast members and original cast member Chevy Chase, who was hosting, for the good-nights at 1:00 a.m., it was my birthday. I was forty years old, older than most people when they left the show; this was just the beginning for me.

A couple of weeks later, I did my first Bill Clinton on air. It was during a sketch about Halloween in New Hampshire during primary season, and all the candidates showed up at someone's door during trick-or-treating hours to pitch themselves. Norm Macdonald did his killer Bob Dole. David Koechner did Phil Graham, and diminutive Cheri Oteri did her hyper Ross Perot. When it was Clinton's turn to ring the bell, I grabbed handfuls of candy and shoved them in my pockets. I had no idea at the time that this was the character that would largely define my time on *SNL*. All I could think was how I was bloated from the lithium, but I guess that made for a convincing pre–South Beach Diet president.

It's hard to say why that character hit so well. I think part of it is that the guy himself is so endearing in his charm and obvious human frailty—thank you, Paula Jones. Clinton was brought back the next week in the cold open. The sketch, written by the current junior senator from Minnesota, Al Franken—I didn't see *that* coming either—had me sitting in the White House kitchen in the middle of the night, stuffing my face with whipped cream from a can and hot-fudge-covered hot dogs while calling people to apologize for being a failure. I could relate, and I guess I got a little carried away during dress. Afterward, Lorne said I was taking too much time with the eating. "This needs to pick up. It's not a one-act play."

Clinton's self-esteem was so low, he asked the pizza delivery guy, played by Tim Meadows, to yell "Live from New York" for him.

I finally got to yell that famous line in the seventh episode, in character as Jesse Helms. I would go on to do it more than sixty times over the years, more than any other cast member in the show's history. (Dana Carvey held that title before me.)

But the first season was pretty tough for me. I had to learn how to be on the show. (I wasn't there long before I was given my own office because I had to practice, and Colin had to write.) What I found really unnerving was that sketches could play well at read-through, get a lot of laughs, and still not make the cut. Since I was new to live television, I had no idea there are a million reasons, technical reasons, why something might not be chosen for air. But time went by, I kept showing up, and I watched other cast members to see if I could pick up any pointers for *SNL* survival.

I was also going to the Comedy Cellar pretty regularly to try out new impressions. Sunday, Monday, Tuesday, I'd go downtown to work on Donahue or Clinton or whomever I was expecting to play in upcoming shows. It was at the Cellar that I first tested out Clinton biting his bottom lip and giving the thumbs-up, and it was a hit. The first time I did it on air, the audience loved it. After that, any time I got a Clinton script, it would include a note, "Does the thumb and lip thing." For the record, I never actually saw Clinton do those at the same time. I made it up.

I didn't really start making it on the show until writer Adam McKay, now one of Will Ferrell's partners along with Chris Henchy and Judd Apatow in Funny or Die, the brilliant comedy video website, took an interest in me. When failed presidential candidate Steve Forbes hosted in April 1996, Adam wrote a sketch in which I played Koppel interviewing Forbes, playing himself, about whether he was the author of an anonymous political novel in which the lead character is a handsome ladies' man named Teve Torbes. At one point, after Forbes, in a fit of giggles, denies yet again that he is the author, I ad-libbed, "C'mon!" Ad-libbing was verboten as a rule—as Lorne has said, it's hard enough putting a show together in six days without introducing sur-prises—but the audience applauded loud enough that I had to pause for a moment before saying my next line. Suddenly I had this half-assed sense: *I think I have a hit character.*

In total that season, I managed to establish a number of impressions good enough to recur: Phil Donahue, Koppel, Clinton, Richard Dreyfuss. When I did Jesse Jackson on "Weekend Update" the first time, Norm Macdonald, who was anchor then, broke up when I yelled at the end of a largely nonsensical rant, "Say it with me: Yabba Dabba Do!" It was one of the few times something I wrote made it into the script and on air.

I even got a little make-out action that first season. I did one cold open as Jay Leno performing for the troops in a USO show. During the bit, I brought out Tim Meadows dressed in drag as RuPaul. Leno starts to say how beautiful RuPaul is, then goes in for a big lip lock. Tim runs offstage in mock horror (at least I think it was mock), and I turn back to the audience, my face covered in his smeared lipstick. In another episode, I played veteran *20/20* anchor Hugh Downs, who ends up pulling coanchor Barbara Walters, played by Cheri Oteri, behind the desk in a hot clinch.

Veterans and newcomers spent all season trying to figure out how to work together as an ensemble. Toward the end of the year, we started having great shows. It took time to turn it around, but suddenly SNL was happening again.

I got a call from someone at NBC saying, "Great year. We look forward to having you back."

I thought, Impossible. I could not have gotten here from where I started.

CHAPTER TWO

The Golden Years

Melbourne, Florida
The 1960s

When I was very little, maybe four or five years old, I used to sit in our little one-story house on Wisteria Drive, overwhelmed with the sense that I was surrounded by evil. When the sun started to go down in the late afternoon, I was filled with foreboding, and everything was scary—the walls, the furniture, the rug, the very air was scary. I'd hear the branches of the hibiscus bush outside the kitchen window *thump thump* against the glass and feel something was coming to get me. Outside, the trees, the grass, the asphalt in the driveway, seemed alive and threatening. I had to make it through the night until our maid Myrtise arrived in the morning.

My mother described Myrtise as looking like a young Cicely Tyson. I only knew her from a small child's perspective: she smelled like the wind and wore sleeveless dresses that showed her arms, which were pretty, surprisingly strong, and—significantly, in the largely racist neighborhood where we lived—brown. I was entranced by the sight of my little white hands in her warm, encompassing brown ones. I knew that we lived among people who used the N-word, but I didn't know that my father's grandfather had belonged to the KKK and ran moonshine back in Georgia during Reconstruction. It was very confusing for me because I adored Myrtise. The only real affection I got during the first few years of my life was from her. How was I was supposed to disparage her at the same time? I only knew that when Myrtise held me, I was safe.

Melbourne in the late 1950s and early '60s was a long way from the colorful playground of Disney World or swank Miami Beach. But Melbourne is one of those places—Florida is filled with them—that's not really anywhere. Squeezed up against the Atlantic Coast, with big brother Orlando to the northwest and the "real" city of Palm Bay just to the south, in those days Melbourne only impersonated a real city. Nevertheless, it somehow manages to pack in a cool 75,000 people, most of whom have jobs and spend their weekends heading out to the beach, or the barrier island that manages to keep the worst of the ocean weather at bay, literally. If you wanted to, you could drive the whole barrier island all the way down Route 1A—it's nearly 100 miles of beach road, dotted here and there with fun things like cuddly Patrick AFB and my personal favorite, Big Starvation Cove.

I guess the highlight is that Melbourne is smack in the middle of the Space Coast, less than an hour south of Cape Canaveral and the Kennedy Space Center, which was established the year I turned seven. That's where the Apollo moon missions, the Hubble Telescope, the space shuttles, and the International Space Station were all launched. Nearby Cocoa Beach is where Major Tony Nelson lived with Jeannie Saturday nights at 8:00 p.m. (If you're under age ninety, you probably didn't get that, but bear with me.)

I think it was thanks to my love for Myrtise that I was always oddly proud of the fact that Melbourne was founded by slaves after the Civil War. And the fact that freed men established my town seemed to offset the fact that it was boringly named after the first postmaster, an avowedly white Brit who actually spent most of his life in Melbourne, Australia. His actual name was Archibald Corncraker McTavish; or maybe it was Tarquin Bartholomew Bungle. No, no: Kevin. That was it. Kevin. Okay, no, it was really Cornthwaite John Hector, and I'm sad that more people aren't named Cornthwaite.

But living in Melbourne was more like living in the Deep South of my ancestors than in the Florida of picture postcards. The people lived lives that were all but prearranged—marriage, mortgage, kids, church—and they all seemed to suffer a kind of lower-middle-class desperation. My parents were no exception.

In those days, especially in the South, if you were a woman with aspirations, you might as well be a whore. My mother, Margaret, who looked a bit like Annette Bening with luxurious auburn hair, was trapped in that world, which dictated that she become a wife and mother. She told me on my wedding day years later that she only married my father, Max, because her own father had threatened "to beat the living daylights" out of her if she didn't.

Wisteria Drive sounds a lot more picturesque than it was. Our house was one of many one-story tract houses on small plots lined up one after the other. Most people tried to keep a lawn, but in the Florida climate most of the lawns were sad, sandy affairs sporting more brown than green. A few people put up white picket fences, but mostly the yards were left to the elements. The Florida East Coast Railway, a 350-mile stretch of freight line between Miami and Jacksonville, ran right behind our house.

My parents kept up appearances and did what they were supposed to do. They had children: first, a little girl, and then, a couple of years later, a son. My dad ran a Western Auto store with his father. My mother belonged to the First United Methodist Church a couple of miles down the road in downtown Melbourne. She made us breakfast in the morning before my sister and I went to school. We ate dinner together as a family every night. My mother talked about Jesus and participated in church functions along with people who made cookies for bake sales, donated clothes for charity drives, helped a neighbor in need. On the surface, everything looked just as it should, very Ozzie and Harriet. (Again, under ninety, sorry.) But that was just for the neighbors.

Before I was old enough for school, each morning my mother and I drove to pick up Myrtise. We had to cross a bridge to get there, and every time we did, my mother would swerve as though she were going to drive the car right off. I was never convinced that she wouldn't do it some day.

On Sundays throughout my childhood, I went to church with my mother, but when I was little I asked questions that got me in trouble. If it was seven days to make the world, how long was a day? Was a day an hour? Was there a sun yet? If there wasn't a sun, how long was a day? One time I made the mistake of referring to the "ghost of Jesus," which was blasphemy.

"He's a *spirit*, not a ghost," somebody said angrily to me. What's the difference, they're both gaseous? At four or five years old, what the hell did I know about the Holy Trinity?

My father didn't go to church much; his religion was the United States of America. Tall and blond, he had a Cary Grant–esque dimple in his chin and shoulders broad enough to fill a doorway. He'd gone to military prep school just as the United States was joining the Allied effort in World War II. Under his junior year photo in his yearbook, Max Hammond had written that it was his dream to go to Duke, play baseball, then go on to law school. This was in the days when baseball was king; aside from boxing, there wasn't much else in professional sports—no football, no basketball, no hockey—so every kid played baseball. My father's high school team was top-notch; one of his teammates was Al "Flip" Rosen, who would become a four-time All-Star for the Cleveland Indians in the 1940s and '50s.

But by the time my father was a senior, the caption under his yearbook photo said his dream was "To serve my country the best way I can." He was sent to Germany an eighteen-year-old second lieutenant, the youngest commissioned officer in the history of the

U.S. military at the time, or so he was told. I never learned all the details of his service, but he came home with a lot of medals and a tortured soul. Armed conflict ruined him.

My father used to tell me about one battle when almost everyone in his company had been killed, and the Germans were looking to shoot survivors. There was a guy lying next to him, the top of his skull blown off. My father reached over and scooped out a handful of brains and smeared it over his own face, and lay there pretending to be dead so the Germans would pass him by.

"You do what you have to do to survive," he'd say.

Once in a while, my father would sit at the dining room table, getting drunk on gin. Staring off into space, he would sometimes begin speaking to his men as if they were going into battle:

Some of you won't be coming home tonight. I might not be coming home tonight. You have to remember that you're part of something that's bigger than you. This is about fighting evil, not someone who's angry or going through a phase or disgruntled, but someone who's crossed the line between sick and bad. They want to stop the country we live in, to stop us from being able to choose the color of the paint on the wall or a career or what clothes you wear. You can be sure that God will be coming into battle with us. God has an interest in what happens here today, and some of us may go with Him. But when these cocksuckers are fucking with us, they're fucking with God.

Over the years, I heard that a lot. He still harbored plans to play professional baseball and go to law school after the war, but then he and my mother got married in Sylvester, Georgia, in 1947, so he had to work to support his wife. Three years later, he was called back for duty in Korea. A week after he shipped out, he was in another war zone, this time in the Far East.

How do you go back to civilian life after that? One minute he was a trained killer, and, as his medals reflect, a damn good one. The next he was supposed to be a hardworking family man, then they sent him back to kill again. He'd lived in a world with bombs and guns and shattered corpses everywhere, and then next thing he knows he's sitting in a lawn chair drinking a martini with our Siberian husky at his side. Sometimes when he sat at the table with that glass of gin in front of him, he would cry, muttering that after what he'd done, he couldn't be loved.

Perhaps that was why whenever I caught his eye, he always looked away quickly.

I don't know if it was his military experience or simply how he was raised back in Georgia, but my dad took shit from no one, and he wasn't shy about telling people what he thought of them.

One day a Jehovah's Witness came to our door. Unfortunately for the poor bastard, my father was the one who answered it. When the guy was done with his glory-of-God spiel, my father looked at him for a long minute before he said, "You're not worth shit, are you, son?" And he slammed the door in the man's face.

On a sun-drenched Saturday afternoon, my father was in the yard when a car came speeding down the road. He stepped out to the curb and waved it down. "You're driving too fast, son. Do it again, I'm gonna bounce ya, and I'm gonna bounce ya good."

More than once I saw him walk up to boys he didn't know and say, "Cut your hair, son." And to one kid who had a ring through his lip, he said, "You didn't turn out too good, did you, boy?"

One night during dinner, my sister told us the veterinarian was threatening to put down her dog because of an unpaid bill. My father got up and went straight to the phone to call him.

"My daughter said you're talking about destroying her dog."

I could hear "blah blah blah" on the other end of phone.

"Well, if you touch the dog, you're gonna get killed, son. Do you understand me?"

More "blah blah blah."

"No, no, this is not about what's right or wrong, what the police will do, or what they can do, or what God would want. This is about what's gonna happen. And what's gonna happen is you're gonna get killed, son."

He scared the fuck outta the guy, and then we ate.

Sometimes he turned his rage on us. We'd be at the dinner table, and my sister or I would inadvertently say something that upset him.

"What in the goddamn hell!" he'd say, then he would get up and walk over to either my bedroom door or my sister's bedroom door and kick in a hole, then walk back to the table and sit down. We lived for years with those holes in our doors. My father bought crude squares of aluminum that he pasted over some of them.

I'd be in my bedroom and hear this crash at the door and my father yelling, "I'm going to bash you halfway through that wall." I believed him.

My father kicked or punched my bedroom door so often, it eventually gave way.

He once said to either my mom or my sister, "Why don't you just kill me? Why don't you just get yourself a gun, put it right up here, temporal lobe, squeeze the trigger, blow my brains out, end my life, get me off of this earth? Because that's where we're going."

My mother was at once more subtle and more sinister. I remember my hand being slammed in car doors and thinking it was my fault. But she wasn't trying to break my fingers; she was saying, "See what I can do? See what I think of you?"

She used to recount cheerfully the time she beat me with her high heel and I began to bleed in front of everyone in the park in Jacksonville.

"I beat you bloody!" she'd say, her hand on her stomach to contain her chuckles.

I used to wake up in the morning wanting a mom. Not *my* mom, but *a* mom. I wanted mothering, the magic I noticed in the hands and the voices of other mothers. Instead, my mother told me that if I ever saw anyone in my room at night, it was ghosts, and I'm not talking about the friendly Casper variety. She was more interested in her church and her friends and playing piano. Her favorite piece of music was Tchaikovsky's Sixth Symphony, which I picked up by watching her and imitating the movement of her fingers.

When I was about seven years old, I finally discovered a way to connect with my mother. I noticed that if I could get her talking about certain people in the neighborhood, she would become enraptured doing impressions of them. She would tell stories and do the voices of Coach Davis, Betsy Whatshername, Maggie Turnbull, and she did them all expertly. Doing my best to copy what she did, I learned to do voices too. If for nothing else, she seemed to love me for that.

We had a recording of Dickens's *Christmas Carol* read by the British actors Ralph Richardson and Paul Scofield, and my mother and I learned all the parts. If I wanted to get her attention, I would say, "Let's do Merry Christmas, Uncle!" We were especially fond of the conversation at the beginning of the story between pre–ghost visitations Ebenezer Scrooge and his rosy-cheeked nephew, which we acted out without the descriptive bits that Dickens had written into the original:

"A merry Christmas, uncle! God save you!"

"Bah! Humbug!"

"Christmas a humbug, uncle! You don't mean that, I am sure?"

"I do. Merry Christmas! What right have you to be merry? What reason have you to be merry? You're poor enough."

"Come, then. What right have you to be dismal? What reason have you to be morose? You're rich enough."

"Bah! Humbug!"

"Don't be cross, uncle."

"What else can I be, when I live in such a world of fools as this? Merry Christmas! Out upon merry Christmas. What's Christmas time to you but a time for paying bills without money; a time for finding yourself a year older, but not an hour richer; a time for balancing your books and having every item in them through a round dozen of months presented dead against you? If I could work my will, every idiot who goes about with 'Merry Christmas' on his lips, should be boiled with his own pudding, and buried with a stake of holly through his heart!"

We used to say that last line all the time, even when we weren't doing the whole scene, just because it was so fun to say. For my money, if you're going to tell someone off, do it in a posh British accent. It's way more effective. I always wanted to use the words "vomitorious" and "oblivion" in the same sentence.

As I got a bit older, I decided to take the entertainment power of impressions farther afield. I started doing Porky Pig saying his trademark line, "That's all, folks!" and Popeye doing his chuckle, which made me an instant star among my friends at school. A pattern for learning new voices quickly emerged: When I listened to a voice I wanted to imitate, for instance a recording of Popeye scat singing, the voice came to me as a color first. In Popeye's case: blue. Then I saw all the letters in my head—*skee da bee dat doh skidibit day id ibit duh bug da yee da doh.* In the same vein, Porky Pig was yellow. Voices still come to me as colors first.

Despite what you're thinking right now—which I'm guessing is something along the lines of, Aha! *That's where he got it!*—I wasn't destined for, or even interested in, a life doing impressions.

What I was really supposed to do was play baseball.

I think all human beings on this earth deserve three golden years like the ones I had beginning at age twelve when I started playing Little League. For me, a tsunami of joy was unleashed with every stroke of the bat. The thrill of hitting a ball well was mind-bending, watching it rise up, the sun glinting off the horsehide, outfielders running back toward the fence. It made the constant sense of foreboding at home fade into the background.

Even the smell of the cigar smoke coming from the wizened old men from the neighborhood who'd come to watch was delicious. They didn't know anybody or have any kids in the league, but it was a baseball game, and, for that night at least, we were all playing for keeps.

And the sound track to the late 1960s was glorious—Motown at its best. People brought their transistor radios to the games. Someone put a turntable in the announcer's booth at Wells Park, and some guy would go up there and call the game. There was a guy named Damon Johnson who brought Motown songs. While we were getting ready for the game, he'd play the Four Tops, the Supremes, Marvin Gaye's "Too Busy Thinking About My Baby," the Zombies' "Time of the Season," the Jackson Five's "Stop! The Love You Save," the Foundations' "Build Me Up Buttercup," "Love Grows (Where My Rosemary Goes)," the one-hit wonder by Edison Lighthouse.

We didn't know that we weren't supposed to love black music any more than we weren't supposed to love Hank Aaron. Some of us white boys used to imitate Hank Aaron's stance batting. To us, he was just a hero baseball player who went to the All-Star game year after year. So was Roberto Clemente of the Pittsburgh Pirates. Whenever I was in a slump, I used to imitate Clemente's teammate Willie Stargell, who hit two of the only four home runs ever to soar out of the confines of Dodger Stadium. It got me charged up so I could hit again.

But above and beyond, New York Yankee Mickey Mantle, one of the greatest players the game has ever seen, was my dude. First Bruce Bochy, then all my friends, called me Mick, because I always wore number 7 in his honor. (Who knew that thirty years later, the Mick would hold a press conference after receiving a liver transplant to replace the one he'd destroyed with booze in which he said, "Don't be like me"?)

Southwest Junior High School integrated while I was in eighth grade. There had been riots in Tampa the previous year, so things were a little tense, not least because a lot of the adults I knew were racist. I remember standing in line in the cafeteria with a forlorn-looking young man whom I tried to engage in conversation about busing and integration. I was probably twelve or thirteen. I was only trying to be interesting. The kid didn't want to talk about it at all. I thought, Great, I'm a fag. You were suspected of being a fag like it was a crime whenever you did something different.

But for a kid, it was an enchanted time. We didn't worry about predators and pedophiles. We went fishing and swimming. We were just little kids running down to the creek, jumping in. Stupid fucking kids.

The creek in Florida? Water moccasins, alligators. But we did it with no fear. Someone had hung a ship rope from an oak tree over the most gorgeous stretch of sun-dappled stream behind the church that ran into the Indian River and out into the Atlantic Ocean. We'd swing out over the stream on the rope and let go, dropping a long way into the golden water below.

If I wasn't playing baseball, I was daydreaming about it. The Little League season lasted only three or four months a year, so the rest of the time we played in the street with tennis balls, or we played in the dirt field at the end of Dunbar Avenue. I made myself a sliding pit there and walked over most evenings at dusk to teach myself how to slide—hook slide, headfirst slide, stand-up slide, any kind of slide I could think of.

When we were out in the street, late in the afternoon and early evening, or in the field at the end of Dunbar Avenue, we would go out of our way to be inventively profane—ball-sucking cock butt-fucking shit sperm cunt-licking fuck-you-up-the-ass cock dick butt motherfucker cock-sucking shit fucking dick cunt fuck—trying to outdo each other in a delirium of cursing. We lived in a very repressed Christian white neighborhood, where we weren't allowed to even say "damn." We laughed our heads off.

Then there were the spitting contests. I don't know if you've noticed, but baseball players spit like crazy. And we wanted to be just like the pros, so we practiced our spitting technique. A lot. I became a first-ballot Hall of Fame spitter. Eventually, I could spit out of all the compass points of my mouth: far left, left, center, right, far right. They were joyous afternoons of spitting and holding forth the most unimaginable tirade of bad language we could summon up.

Despite what our parents and the church told us, the earth didn't open up and swallow us whole.

For some reason, there were a bunch of older kids in eighth grade—fifteen- and sixteen-year-olds who'd been to reform school—and they were *rough*. During phys ed classes, the coaches had to figure out a way to keep them from going AWOL, so they put them all in one class with us thirteen-year-olds—we were frightened out of our wits. One day when were playing softball, this man-kid named George Robinson hit an infield fly that I barely had to run to catch.

After the game, he came up to me and said, "Don't you ever—*ever*—do that again."

I assured him that I would not.

One kid named Angus Verandah was like twenty years old. He'd been in major scrapes and was considered violent. Apparently, he thought I had seen him do something and had fingered him as the culprit at a school event, which caused him trouble with his parole officer. Kids in school told me this guy was going to beat me up, or cut me. He was a real kingpin, so I knew where he hung out. For two days, I avoided him. Then I decided I'd rather be cut than run from this guy anymore.

"I hear you been looking for me," I said, trying to muster some bravado in my squeaky white voice.

"I heard you fucked up my shit," he said with a voice so deep it made my sneakers vibrate.

"I don't know you. I don't know what information you're getting, but it's not right." I noticed I was surrounded by a group of his friends now. "If you've got to do something, I understand."

He looked in my eyes to see if there was any mendacity there. When he didn't detect any, he walked away. Then his friends walked away, and it was over. That's when I realized that in life there are moments of truth for all of us, and there is no getting away from them. Sucks, right?

I used to go out in my yard at night, with a streetlight casting my shadow on the wall of our house, and practice my swing. My teammate Bruce Bochy and I used to go out in the field sometimes and he'd pitch to me, just me and him, in our bare feet. You could get lost hitting baseballs in the summer heat. My best friend Wayne Tyson pitched to me endlessly.

That was a glorious season. I would be out there sweating, spitting, cursing, laughing, getting sunburned, loving it when I had to dive for a ball. It was a heady, wonderful experience. I hit .334 that year. The year before, we'd been on a Little League field, but now we were playing on a major league-size infield. I could barely lift the fucking bat. I did a one-hop one off the fence that year, but the fence was only 300 in the center and 297 down the line.

Bochy was already functioning at a higher level. He was small then, but he had a major league arm. The first time my dad saw him throw the ball, he stood up like a shot and said "Jesus H!"

Twelve was good, but thirteen and fourteen were better than anything that ever happened. Nothing but sweating, body surfing to cool off, sweating, swinging, sweating, swinging, sweating all day. Some of the folks in the neighborhood would let us use their garden hoses to cool off. Heavy calluses formed on our hands. Every day, the second I woke up, I wanted a bat in my hand. I couldn't wait to get my hands around the bat handle, go out in the neighborhood, and rustle up the kids for another game with a tennis ball. Drag bunting, sliding, everything. We were entranced.

There'd be two games per day in Babe Ruth League. Even if your team wasn't playing, after you jumped in the ocean to cool off, you went to the games. And to the people of our little community in the stands, those games were as important as any major league game. A lot of the guys on our team were surfers. That's what that community was: football players and surfers. On the inland side, you had the baseball players. The other guys would surf all day before the games. Both sides would come together at the baseball games and we would play like it was the fucking World Series. Every swing of the bat was the same as if it were Mickey Mantle at the plate. Sometimes before going out to do an *SNL* sketch, I'd think about the two doubles I hit off of C. P. Yarborough in one night. If you were a baseball player in Melbourne, that was one of the big things that happened to you: you were going to bat off this guy. He was one of a number of fabulous athletes I knew who demonstrated tremendous prowess, then lost interest and went off to do something else.

For the rest of my life, in the most dire circumstances, I would think of certain base hits to remind myself of a time when all was well.

During my freshman year in high school, 1969–70, women decided it was okay not to wear bras. You're a thirteen- or fourteen-year-old boy, and now there are nipples everywhere. These were all beach girls. No matter what color their hair really was, the sun transformed it—in brunettes there would be lovely auburn streaks, girls with brown hair would have these startling natural streaks of blond. Everyone had deep tans. These were girls we'd known since we were five, and now I couldn't make eye contact. I'd get so weak I had to sit down. If a girl brushed against you in the lunch line, you could burst into a sheet of flame. There was a chick named Lisa Kasarda I'd known for years, and now she wasn't wearing a bra. How was I supposed to talk to her?

Veronica Baxter was a year or two older than me, and she was the loveliest thing—human, mammal, flower, anything—I had ever seen. And I wasn't the only boy who felt that way. When she walked past, I would have to stop what I was doing, walk over to a bench, sit down, and put my head in my hands. *My God.* Veronica used to dress in a kind of European style, so she showed cleavage. The first time I saw her cleavage was the only time in my life I said Jackie Gleason's words, "Hommina hommina hommina." She took me for a ride in her European sports car once. I never intended to try anything with her at all, I was too intimidated by anything with that majesty and power and beauty. Also, she had a boyfriend who was a mastodon. I'm *still* afraid of him.

And the guilt helped. We were all going to church every Sunday, so we knew we'd go to hell if we partook of what we were seeing if through some outlandish stroke of luck we should be alone in the same room with a Lisa Kasarda or a Veronica Baxter. Clearly God didn't want that. In fact, I was convinced the whole world was going to hell because those girls couldn't be allowed to show their breasts like that without God intervening.

During that golden summer before my sophomore year at Melbourne High, I hit .500 and lost my virginity on the beach to Miriam Bonaparte. She was a fuckin' knockout.

There wasn't a lot of discussion about the facts of life in my house or anywhere else. It seemed best to everyone, including my parents and friends, if sex didn't exist at all. My mom's best friend Theresa had a male dog, and one day he mounted a girl dog from the neighborhood. Theresa came running from the house, screaming and clapping her hands together, as if a murder was being committed.

That was the kind of world I was in. But two four-teen-year-old kids left by themselves, not quite under-standing, don't need any more help than what Mother Nature and DNA will provide. When I finally lumbered into a sexual relationship with Miriam, my powers as a human—my power of speech, my reasoning center, my executive command center—all that shit was out.

I did understand that babies came from this, and I was horrified, but far worse than any concern about unwanted pregnancy was the guilt. I thought some-thing horrible would happen to me, that I would go to hell because I could not stop having relations with the stunning Miriam Bonaparte. At that age, you have lots of interest and lots of energy, but I felt like a sinner the whole time.

After I had sex, I was never sane again.

That same summer, some friends of mine miracu-lously got their hands on a six-pack of Busch beer. We sat down in the ditch between Parsons Avenue and Dunbar, Australian pines covering us to the right, paper trees covering us to the left, a stream trickling beside us. I drank two Busch beers, and the world changed—no longer a harsh and lonesome windswept prairie, but a lush fertile valley bursting with luscious fruits and spring lambs rollicking in the meadows.

Like something out of a Disney cartoon or Dorothy landing in Oz, the universe went from black and white to color. I half expected a little blue bird to land on my shoulder and start singing or a band of orange-hued little people to escort me to Emerald City.

That shit was good.

As I went into my sophomore year of high school, I tried out for the junior varsity football team, even though I'd never played football. My father really wanted me to play, so he was pleased when I made starting quarterback. But rather than teach actual tactical skills of combat within the rules of a contact sport, the coaches focused on teaching us how to be brutal.

The coaches used to have us line up across from our best friends and take turns slapping each other in the face six times. Everyone laughed about it at first. You strike the guy mildly, and he laughs. But he strikes you back harder, and you laugh a little less. You strike him back harder than he struck you, and he doesn't laugh now. He strikes you hard. You hit him back harder. That's the natural human instinct.

There was this one kid named Johnny, fifteen years old, six foot four, a fabulous athlete. But he also surfed, and to some people on the inland side of the Indian River in Melbourne, surfer boys were reprehensible, or even worse, gay. One day a coach was slapping Johnny on the helmet while he ran in place.

"Whatchoo doin', surfer boy? C'mon, surfer boy." Then leveling the ultimate accusation, "I don't think you want it." It was the greatest indictment that could be leveled against a player, that he didn't want it.

Johnny stopped running and said, "You know what? You're right. I don't want it, okay?" And he left the field.

As punishment for "not wanting it enough," the coaches put us in a ditch thick with red and yellow glass-cutting sandspurs to fight each other. We'd get those spurs all up in our helmets and practice jerseys. I don't think I'd ever been in a fight before, except once when I was really young and I boxed with another little kid; we wore professional boxing gloves that were larger than our heads, so we did little harm. I was afraid that I wasn't a man because I didn't want it, so I got in the ditch with the third-string quarterback. I dropped the guy with a quick blow, and something inside of me convulsed.

The coaches started yelling, "Get you some! C'mon, Darrell, get you some!" exhorting me to continue to strike a man who was down. I hit him one more time when he was down. I've never hit anyone since then. I can't tell you how much shame I felt. It really isn't sport if the other guy is already down. It certainly isn't football. It reminds me of the coach from Eau Gallie High School who used to bite the heads off live frogs in the interest of getting his players fired up. The bloodlust of hurting someone who was down was supposed to toughen us up.

The slap drills ended when my friend Frank Facciobene got a punctured eardrum.

In junior year, we got a new coach. At practice one day, while I was calling plays in the huddle—"Hut One! Hut Two!"—he said, "You gotta say it faster."

I said, "Like H-T, instead of H-U-T?"

I might as well have dropped to my knees, pointed to his crotch, and pointed to my mouth. "I wouldn't know about that."

With so much sin in my life, I was a haunted vessel, a satellite of the Dark One. I first got drunk on a Saturday night at the end of the summer before school started. I figured I was already ruined, since I'd been having sex. How much worse could it get?

My friend Ernie from the football team—a real tough kid—called me up one afternoon. "Come over to my house. We'll get drunk, man. My parents are out of town."

So we took two eight-ounce highball glasses, filled them roughly three-quarters of the way up with

Seagram's Seven, and added a little 7UP. And we drank them down. We had no idea we were handling nitro-glycerin. The next thing I (vaguely) recall, we were slow-dancing with each other.

I was supposed to stay over at his house, but I ended up walking the five miles back to my house, vomiting the whole way. My parents came home from a party not long after me, and they came into my bedroom. My mother was trying to have a conversation with me, but I was unable to have that conversation. My father suddenly said, "Ohhhhhhh. Margaret, c'mon, let's go to bed. Leave sombitch alone."

The next morning, I was in church with my mother, desperately hung over and full of evil. I'd already had sex, and now I had gotten drunk, and I had slow-danced with my teammate. I was waiting for the gates of hell to open right there in the church, the winds to pick up, and the preacher to point his finger at me: "YOU!"

When nothing happened, I was even more confused. I knew guys were smoking pot and drinking and having sex, and none of us were going to hell. It threw me forever after. I was always nagged by the terrible reality that God had not shown Himself when I had sex, that God had not shown up when I drank two beers. The week after I drank all the Seven and 7s with Ernie and ended up slow-dancing with him, I hit a magnificent double in Eau Gallie. *Is God here? Does he care? Is he paying attention?*

Then I thought, Maybe there is no God, and the whole structure started falling down for me.

And then my mom caught me masturbating. As was her custom, she didn't knock on the door, she just barged in. Since no one had explained to me what was happening to my body, and why my hands wanted to wander there, I was fumbling my way through that process as well. I was mortified.

My mother's reaction was silent rage. In the days that followed, I'd walk up to her and try to have a conversation, and she'd behave like I wasn't in the room at all. Finally, I grabbed her by the arm and said, "Why won't you talk to me?"

"Why don't you go into your room and pee on yourself?" she spat back.

I thought, Wow, this sex thing is really gonna work out good, huh?

Not long after this, my mother came up to me one day and, for no reason I could fathom, started hitting me, really trying to knock me off my feet. I was fourteen years old, and I was a good athlete, strong. I was probably four or five inches taller than her. I put my arms out in front of me and let her swing away at me in her blind fury. She kept swinging and slapping, and then, too tired to raise her arms anymore, she stopped and looked at the ground. That was the end of it.

I had reached puberty and was changing into a full-blown fur-bearing sexual creature, and it had made me an object of scorn. I had every reason to believe that I was a malignant and terrible person for allowing this to happen to my own body, as if I were somehow responsible for it all.

CHAPTER THREE

There's Something Wrong Here

When I was still in grade school, my grandfather decided to pull his money out of the Western Auto business. My father couldn't keep it going on his own, and the store went bankrupt. Ironically, my sister and parents and I had to move in with my paternal grandparents on the western side of Florida in Tampa while my father found work as a traveling sporting goods salesman. After a year or two, he was able to move us back to our house in Melbourne—they'd rented it out while we were gone—so from then on my father was on the road a lot.

But he took a great interest in my young baseball career. He came to as many of my Little League games as he could. The second year I played, I graduated to Babe Ruth League, and my team had the great misfortune of being sponsored by Howard's Meats. (Go ahead, laugh. Nobody can laugh about that name more than a bunch of thirteen-year-old boys did.) My dad became assistant coach.

My father and I became obsessed with the 297 sign in the outfield at Wells Park. One day during the winter when nobody was there, he took me out to the field and stood at the plate, explaining why a hanging curveball was important if I wanted to get beyond that sign.

"If the pitcher puts his finger on the wrong seam, throws a two-seam curveball instead of a four-seam curveball, and that ball hangs in the air just for a second before it breaks down to the plate," and he pointed to the 297 sign and, with missionary zeal, said, "Buddy, that sumbitch [pause for effect] is *gone*."

It was one of the most exciting afternoons of my life. None of the boys could hit a ball over that fucking fence, but he was telling me I could. After years of being terrified of him, it was nice to have him on my side about something.

"Some time in your life, someone's gonna hang you one right up there, and, buddy, it's gone." It was like Billy Graham saying, "Mine eyes have seen the coming of the Lord!"

Throughout his life, my dad had this charming habit of standing up whenever something in a game excited him, even if he was just watching the Florida Gators on his TV at home. So I knew I was doing well when he stood up during one of my games.

But my father being my father, there was a downside to his being so involved in my baseball world. During one game, there was an argument with the umpire, and my father must have had some Hitler flashback, because he went to the opposing coach, in front of the entire community, put his finger in his chest, and threatened to kick the guy's ass.

When I was in high school, the opposing coach became the varsity baseball coach. From then on, if I didn't get a hit my first time up, the coach would bench me. I didn't handle the rejection very well, and I started to bat terribly.

Part of the problem was playing football. I played football until my junior year, when I threw my arm out due to improper weightlifting technique. I also got knocked out three times. In boxing they say a boxer has a glass jaw, which means if you pop him on the chin, he'll go down. I have a glass head. If you hit me on the top of the head, you knock me out. By the time I was done, I had done permanent damage to my throwing arm as well as my back and knees.

Another part of the problem was my drinking. My buddies and I drank together a fair amount, but I had also started drinking alone. I didn't know I was suffering from depression, or where it came from, I just knew I had this great need to escape myself. I started saving my lunch money and going to this store where this zombie who appeared to have been recently unearthed, and who always seemed to have a dab of mayonnaise in the corner of his mouth, would sell us booze. I would sit in my closet in the dark with a bottle of vodka, and if I drank enough, I could imagine that I wasn't part of my world. By the time I got to be fourteen or fifteen, I drank almost every day.

I was still trying to play baseball. I fully intended to make the major leagues. Even though I was drinking too much to really perform the way I could have, I was still good, if unreliably so. When I was sixteen, I finally conquered the 297 sign. In fact, there was a slab of concrete about eighteen feet long beyond the sign, and I was able to hit a ball well onto that concrete.

Then they built this gorgeous new field out in the forest where we played American Legion ball. Nearly three-quarters of all college players played Legion ball in high school, and more than half of all major leaguers did as well, including Yogi Berra, Ted Williams, and current New York Yankees first baseman Mark Teixeira. More than fifty Legion players ended up in the Baseball Hall of Fame. So, you know, no pressure.

Anyway, the fence in left, for reasons that no one could quite understand, was 340. One day, I hit a ball over that fence. My father was at that game, and he paced it off to 420.

My father hit me for the first and only time that year. When I got depressed, which happened more and more often then, he would say, "You got yourself sulled up," which is "sullen" in normal people parlance. I guess I'd finally gotten sulled up once too often, and it pissed him off. Before I knew what was happening, he popped me a shot to the face.

Not long after that, I came home drunk one night, and my father started yelling at me in front of my friends. I was pretty big then, but still, they don't call it "liquid courage" for nothing. I got up in his face.

"My whole life you've been telling me you're gonna knock me halfway through that wall. Do it. Let's go."

After threatening me all those years, now he simply looked at me and walked away.

My batting had become so bad that I struck out the first twelve times I was up during my senior year. My father stopped coming to the games because I was striking out every single time I got up. But on my thirteenth at bat, against Cocoa Beach, I hit a fabulous home run at Wells Park. It was a fantastic relief, but it wasn't to last.

I tried out for the baseball team with my friends at Brevard Community College. I hit really well during fall practice, and I made the starting lineup for the spring. This was no small feat. A bunch of the guys on the team would be recruited by the majors: Greg Fairbanks was drafted by the Texas Rangers; Jeff Simons and Marshall Harper by the Chicago White Sox; and Mark Van Bever by the Phillies.

And Bruce Bochy, our catcher, was drafted in the first round in 1975 by the Houston Astros. After playing for the Astros, the San Diego Padres, and the New York Mets, he became a manager, spending twelve years with the Padres, earning them a National League pennant. In 2010 he took the San Francisco Giants all the way to the World Series—and won. (In high school and junior college, a bunch of us would go dancing at local nightclubs. We all imitated his dance, which we called "The Boche." To all you Giants players, ask the Skip how it goes, it's a real panty dropper.)

Meanwhile, I went to a couple of different radio stations in Melbourne with little sports shows that I'd record for free. I interviewed players in the area, including members of my own team. The broadcasts ran in the middle of the night, but the interviews were pretty good—I knew what it was to be a jock, I had access, and I was free, so they let me go on the air.

The first day of the season was at Twins Field in Melbourne, where the Minnesota Twins farm club trained. It was one of those crazy days that any hitter knows about, when you know you're going to make something happen that day. You can feel it. My father came and threw a few balls to me before the game, and I was popping them out of the park. I even hit a line drive that glanced off his head.

I got up in the first inning and walked. The guy didn't throw me anything close to the plate.

And then it started raining.

The next day I woke up with a fever. I stayed sick for three weeks. I thought God was finally punishing me for having had sex with Miriam Bonaparte.

In my place, the coach put in my friend Wayne Tyson. Wayne and I were batting practice buddies, but he hadn't made the original starting lineup. Now he was spectacular. He hit from the left side, spraying the ball all over the place. He had super speed to first base.

He kept my spot for the rest of the season. It made things awkward between us, and we didn't hang out much anymore.

I still played from time to time that year. In a game against Seminole Community College, I saw my first slider. Greg Fairbanks had a ninety-something-mile-an-hour fastball, and I could hit that, but I couldn't hit a slider. I couldn't hit a curveball that looked like a fastball. Either you are as fast as a cat chasing a feather, or you're not. George Brett was. Bochy was. Marshall Harper was.

That's when I realized I wasn't going to play major league baseball.

I tried to play the following year, but with that slider in my mind, my heart wasn't in it. I made the team, but I didn't hit well, so I failed to make the starting lineup. If I wasn't going to make the major leagues, then there wasn't anything in it for me.

About midway through the season in my sophomore year, I quit.

Soon after, while at a school party, I cut myself for the first time with a serrated steak knife I'd found in the kitchen of the house where we were. I made the cut on the side of my wrist. It wasn't just my failed baseball career, but this overwhelming sense that something major was wrong with me that prompted me to do it. It was my way of telling the world.

Three of my good friends carried me upstairs to the bathroom and cleaned me up. No one ever spoke of it again.

The next time I cut myself, I was alone at home, and I targeted my knee. It bled so badly it wouldn't stop. My father came home unexpectedly to find the kitchen floor red with my blood.

He looked at me and said, "You're sick."

But he, too, never mentioned it again.

On another occasion when I cut myself, my sister took me to the emergency room. I told the doctor that I had been in a fight. He looked at the cut and said, "You weren't in a fight, that's surgical." It was the first time someone looked at me with a mixture of pity and revulsion, as if to say, *There's really something wrong with you.*

Most people experience moments of terror at some point in their lives. Maybe you were on a plane and you hit an air pocket and the plane dropped for a second. Maybe you had a near miss on the freeway. One day when I about ten years old, I went fishing with my father in a small boat in a nearby swamp. As we were nearing a tuft of land, my father screamed at something behind me. I turned and looked, and there was this enormous cottonmouth snake coiled up, ready to strike.

Now take that instant and prolong it—not for a night, not for a week, not for a month, not for a year, but for *decades*. That was the kind of fear that overtook me the moment I arrived at the University of Florida in Gainesville. I had no idea why, but I was in a state of terror that first day. It was perhaps ironic that it struck when I was finally out of my parents' house. It was so bad that I had trouble making words come out when I spoke, like one of those nightmares where someone is coming to get you and you scream but nothing comes out. Even weirder, if someone spoke to me, I could see their lips moving, but I didn't hear the words until after they'd stopped talking.

I had rented a place at the Picadilly Apartments, known for its proximity to campus. When I woke up the second morning, I looked out my kitchen window at the traffic down below. I remember thinking, How can they just sit there in their cars in all that traffic when this is happening to me here? And as I watched the cars stop and start, I realized I couldn't hear anything. I walked out of my apartment and stood in the hallway, and I couldn't hear the traffic. No honking, no engines, nothing.

I went to the university's mental health clinic. The poor grad student who interviewed me grew increasingly nervous as I tried to explain what was happening to me. After a time, she excused herself and brought in her supervisor. It was the first, but not the last, time someone said to me, "We don't have what you need. We're not able to take care of you here." But that didn't stop them from giving me a shitload of antidepressants and an antipsychotic prescription for good measure— Triavil, Elavil, and Mellaril. Sounds like the members of a 1950s girl group, but the only number this trio did was on my brain.

Nevertheless, I went ahead and enrolled in classes, hoping the fear would subside.

One afternoon an attractive coed came up to me, and said, "Hey, my name is Betty-Ann. I'm in your journalism class. Some of us are going out for beers later, would you like to come?"

I opened my mouth to answer her, but I felt as though my face had been shot up with Novocain. I could see her puzzlement turn to alarm, and she ended the conversation.

So I did the next logical thing and signed up for an acting class. Oral interpretation, to be specific. The professor's thing was to help students learn how to project to the back of the theater. He wouldn't cast you in the plays he directed if you didn't have a decent voice.

I started to do a scene one day in his class, and I couldn't get words out. He interrupted me and said, "If you were underwater and had a snorkel and flippers, it would sound like that."

Everyone started laughing at me. Needless to say, I was never cast in any of his plays.

Then I auditioned for a play called *When You Comin' Back, Red Ryder?* which was written by Mark Medoff, the same dude who wrote the classic *Children of a Lesser God.* Let's put it this way: like *Children of a Lesser God, Red Ryder* ain't exactly a musical. When I got the part of the cook, Stephen—Red Ryder's his nickname—it was a new play; it had opened in New York just a couple of years earlier. My character slings hash at a diner in New Mexico; some time before the play begins, someone has decided it would be fun to build a new highway around the place, so it's down on its luck. I banter with Lyle from the filling station next door, and then there's the quiet little waitress (her name's Angel, of course), and my boss, Clark, who's kind of gruff.

Eventually two couples arrive—the well-heeled Richard and Clarisse, who are waiting while their big car is gassed up; and Teddy, a fucked-up Vietnam vet, and Cheryl, his hippie squeeze. In Act 2, Teddy gets all weird and shit and holds the place hostage, making us all do stuff we wouldn't normally do. He leaves, finally, thank God, but his games make Stephen decide to leave New Mexico and head east. I guess that meant something to me, somewhere deep down. I mean, I didn't want some psychotic Vietnam dude forcing me to New York, but I guess when he says, "You never would've found the real Red Ryder sittin' around a dump like this starin' at some tourist lady's tits," he had a point. (Not that I'm against tits, you understand.)

I was most fond of the moment in the play near the end where Stephen gets to tell Clark—after he admonishes him for drinking coffee and "greasin' up the customer's newpaper"—"up your hole with a ten-foot pole!" It's something I'd said often and with great glee when I was a kid.

Everyone knew that I was a fucking joke because I couldn't really talk. But Stephen is this punk who's very inarticulate and can't express himself very well. He's all mouth and bluster, and he tries very hard to be a man of substance and valor, when he's this terrified little guy. I got the part.

After the first rehearsal, the director came up to me and said, "Are you okay?"

"Yes," I lied. "I think so."

"How can you do that, what you just did?"

"I don't know."

"They can't teach that. I can't teach that. Are you sure you're okay?"

A few weeks after the play wrapped, the director wrote to me. In the letter, he said, "I strongly suspect you are an authentic genius." The truth of the matter was, I was playing a scared guy, and I was scared. It showed.

But all of a sudden, I'm a genius. The girls started coming around. "Ooh, look at this blond kid. He's a genius, he can't talk. He's like Marlon Brando." They called me a mumble scratch actor, which is how Brando and Dean had been described.

But I doubt Brando and Dean were drunk all the time. I didn't drink before a performance or a rehearsal, but I was drunk every night, and I was drunk most mornings. That seemed to fix the problem.

Meanwhile, I was working at the university radio station. I spent a year alongside the Alachua County Suicide Prevention Center on a nineteen-part series on suicide prevention. I interviewed students who had attempted to kill themselves as well as health professionals around the state commenting on why people do it. A congressman was going to hold a press conference about it, but it was cancelled at the last minute when the spots were deemed too disturbing to air.

And then a revelation: Truman Capote spoke to us in front of the library. Microphones had been set up in the audience so students could ask him questions. The first question, from a kid with a distinctly Southern husky voice, was, "Are you a homosexual?"

Capote paused, and then in his tinny fey voice said, "Is that a proposition?"

Four hundred gay-hating white kids went insane laughing.

I thought, Whatever that is, I want that.

CHAPTER FOUR

From Hell to Hell's Kitchen

New York City
1977–1995

I moved to New York after college with my friend Larry to try my luck as an actor. Our basement apartment was at 340 East Eighty-fifth Street between First and Second avenues. We were so poor we once went two days without eating. If we made $5, we bought a tuna sandwich and a couple of beers. One night, we heard Martin Luther King's voice booming from a neighbor's radio across the alley. It was a brilliantly edited version of "I Have a Dream," followed by "How Long, Not Long," the speech he gave after the march from Selma to Montgomery. We both wept.

I was working on Fifty-second Street at a place called the Boss Pub, later called the Boss Saloon. I'd heard that you could see everything that had ever been on TV at the Museum of Broadcasting, which was a few doors down, so between shifts I'd go watch Martin Luther King.

By this time I'd known something was wrong with me for a few years, but no one had been able to tell me what it was. When I heard this line in "I Have a Dream": "With this faith, we will be able to hew out of the mountain of despair a stone of hope," I believed, if only for a moment, that I would find some answers. That was part of what the cutting was about, that someone would notice and say, "I know why you do that, and here's the solution."

I had stayed drunk through college, and I stayed drunk in New York. I used to walk around Times Square at all hours of the night, carrying a quart bottle of Budweiser.

Thirty years ago, the "Crossroads of the World" was not the family-friendly mecca of Disney, MTV, the Lion King, double-decker tour buses, swank hotels staffed by German runway models, pedestrian plazas, and outdoor cafés serving double chai lattes. Hell, these days they advertise walking tours of Times Square *you take at night*. In the 1970s, you took your life into your hands if you were there after dark. It was the putrid cesspool of *The French Connection* and *Taxi Driver:* peep shows, strip clubs, triple-X movie houses, and head shops that sold tourist T-shirts, bongs, and fake IDs. Prostitutes for every appetite trolled Seventh Avenue looking for prospects: women with 48JJ breasts spilling from fake rabbit fur halter tops and lipstick red hot pants, six-foot-tall Puerto Rican transvestites in stilettos and fishnet body stockings, toothless old heroin scags charging $2 extra to take their dentures out for a blow job, young boy runaways in gold lamé and fake eyelashes making eyes at the wizened chicken hawks before returning at dawn to sleep in the Salt Mines on the West Side Highway, where the Sanitation Department kept the salt for snowy winter roads. The pimps were a spectacle themselves: one guy staked out the corner of Forty-sixth Street and Broadway, sitting in an easy chair and resplendent in a full-length fox fur coat and matching hat, a gold-and-white telephone, with a wire running to a nearby utility pole, by his side. And my personal favorite, a guy with a sign that said, TELL ME OFF FOR A DOLLAR. In other words, *Good Morning America* did not greet the nation from a street-view glassed-in studio every morning with a live audience peering in from the corner of Forty-fourth and Broadway. Middle America was not ready for *that* tableau.

And here was this naive drunk white kid from the South wearing a black T-shirt that read, "How Long? Not Long," across the front, walking among the derelicts and reprobates, the dregs and untouchables, proclaiming Dr. King's words as loud as I could.

No one even looked at me.

(Flash forward twenty years, I'm drunk as fuck, doing a show at one of the big colleges in Milwaukee, and I'm still pining for freedom from my inner torment. At the end of my set, I say, "I don't want to do any more comedy. I'm going to close with the last ninety seconds of 'I Have a Dream.'" I do it in King's voice, but I'm deadly serious. The audience is all white kids, and the place goes nuts. They start clapping before I even finish. When I call out the last line—"Free at last! Free at last! Thank God Almighty, we are free at last"—standing ovation from a hall filled with white kids. It wasn't for me, it was for Dr. King.)

I wasn't intentionally pursuing a comedy career then—serious theater was in my head—so it was only inadvertently that I did my first New York stand-up in Hell's Kitchen, an area on the west side of midtown Manhattan. A neighborhood doesn't get a name like that without a reason. A gritty industrial area that sprang up when refugees of the Irish potato famine poured into it in the middle of the nineteenth century, Hell's Kitchen was an overcrowded, crime-ridden collection of tenements that became particularly violent during Prohibition in the 1920s. Fifty years later, the neighborhood was still populated largely by blue-collar Irish Americans with a healthy dose of Puerto Ricans thrown into the mix. *West Side Story,* anyone?

To pay the rent, I worked in the kitchen at the Skyline Motor Inn on Tenth Avenue at Forty-ninth Street. These days, the newly renovated Skyline Hotel boasts about its heated indoor pool on the penthouse floor with great views of the Manhattan skyline (get it?), but twenty-five years ago the bar was where gangsters had sit-downs. At least that's what I read years later. At the time, I didn't know fuck about it. On occasion, the manager stuck me out front as a substitute bartender.

One day a guy with dark hair and a pretty fierce mustache came in and ordered a spritzer. When I put the drink down in front of him, I told him what he owed. He looked up slowly and glared. "I don't pay," he said.

I went into the kitchen to find the manager.

"This guy at the bar says he won't pay," I said.

The manager pushed the door open a crack and peeked out.

"Oh, hell, kid, that's Jimmy Mac." The name meant nothing to me, but it obviously meant something to my boss. "You don't want to fuck with him. He only pays for every fourth or fifth drink. But don't worry, he always leaves a huge tip."

He was right. I did well that night.

Most of the clientele were guys I thought were knock-around longshoremen; I didn't know that most of the guys who worked the West Side piers during the post-war boom had been forced to find other ways to make a living in the wake of the introduction of container shipping in the 1950s. No matter, they were good guys. Sometimes I pulled out my Porky Pig impressions to entertain them. They started egging me on, encouraging me to do my impression of Rodney Dangerfield.

"Why are you working behind this bar?" they'd say. "You should be out there making money doing impressions."

These tough dudes had me believing that I could make people laugh. One night they encouraged me to do a set at Catch a Rising Star, and I decided to give it a shot. I went on at 1:30 in the morning and did ninety seconds—Porky Pig, Peewee Herman, and Sylvester Stallone doing Shakespeare: "Yo, Juliet, you know, in the moonlight, I can hardly see your mustache." Three of those same guys came and cheered. I liked the feeling.

They invited me to go drinking with them a few times to a place called Teabags on Tenth Avenue. One night when I was there, one of the guys said to me, "Whatever you do, Darrell, don't hit on a girl here."

I wasn't in the habit of doing that anyway, but out of curiosity I asked, "Why?"

"Every person here has a gun," he said.

Where I'm from in the South, everyone had a gun. What's the big deal? I didn't think any more about it. It would be fifteen years before I learned the truth.

When I wasn't slinging burgers and gin, I got cast in a few plays. And these weren't just any plays. I was going up against New York competition, which is a pretty good indication that you're a decent actor. I did a few off-Broadway productions, which were great. But more often, I landed out-of-town summer stock plays upstate or in Rhode Island. Those were great, too, but each time I got one, I'd have to give up my apartment and my job, which was a huge pain in the ass when I came back to town in the fall and had to start all over again.

Also, I was drinking. A lot.

Sometimes the craving was worse than any desire for food or water or sex. When it hit, all of a sudden—bang— any plans for the next few days were off the rails. *That super interesting woman who agreed to have dinner with me tonight? Not gonna happen.* I remember walking up

First Avenue one morning with a quart bottle of Tanqueray gin in my hand, and I saw Jackie Onassis walking toward me, and all I was thinking was that I'd have to cancel my date for that night because I had the craving.

Finally, about five years into my attempt at making it in the big city, I realized I was sick of myself and my life. I said, Fuck it. I was done with New York and acting.

At twenty-seven years old, with no money and no prospects, I moved back in with my parents in Melbourne. They must have been so proud.

Back in Florida, I found a job at an automated radio station in Vero Beach pushing buttons three times an hour. That gave me access to their production studios, which I used six nights a week for well over a year to practice voices of cartoon characters, sitcom stars, celebrities of every stripe. Given how much people seemed to like them, I thought I would see if I could expand my library.

I didn't think I had as much talent as other people, so I figured I would have to work harder than everyone else. I'd be sitting there in the middle of the night working on my Porky Pig and saying to myself, "My competition cannot be doing this. They cannot be working this hard."

There was one other perk to working there: one of the producers told me that pitcher Bob Feller, the first American Legion alum elected to the Hall of Fame with the most wins in Cleveland Indians history, would pitch to members of the media. Even though I was a button-pusher, I qualified, so I made a beeline for Holman Stadium. When it was my turn at bat, I one-hopped one off the wall in left field.

Feller said, "You got a nice swing, son."

I didn't want to fuck things up, and I was afraid my drinking would do just that, so I went to my first rehab. A doctor at Heritage Health Clinic in Sebastian, Florida, decided I was manic-depressive and schizophrenic, which seemed to make my parents rather happy. It reassured them that my drinking was no fault of theirs. (Other doctors would later tell me I was bipolar or had borderline personality disorder, and even multiple personality disorder. It would be nearly twenty years before an ER doctor in New York would finally make the right diagnosis. But I'll get to that.)

When I got out of the Heritage Health Clinic, I made a decision that I was either going to starve to death, or I was going to do voices for a living.

I heard that a radio station in Orlando, BJ105—Howard's Meats to BJ105, my life was turning into one long dick joke—was looking for a guy who could do voices. This was the advent of the "morning zoo" concept in radio, which Scott Shannon started in Tampa. They always had guys who could do voices on those programs, so I sent them a tape. By then, I was sharp. It was the most I'd ever improved in anything in my life, and I got the gig. Then I started doing voices for radio stations in Phoenix and Detroit. I had a nice apartment in Winter Park. I had a nice car. I was going to AA meetings and keeping clean. I even sponsored someone in AA, a young lawyer named Dean, and we became friends. Life was pretty good.

I did my first stand-up at a club called Bonkerz, owned by Joe and John Sanfelippo, who later became my managers for a brief period. It was an open mike night, and I had three minutes. I did impressions of Pee-wee Herman, Eddie Murphy, Porky Pig, Popeye, Elmer Fudd, Donald Duck. The sound of strangers laughing triggered something in my already addiction-prone psyche. I was hooked.

The other thing I did when I got out of detox was go to a party with some of my old friends and drink. I told everyone that I'd been in the hospital for stomach problems, gastroenteritis or some shit.

Frank Facciobene said, "You didn't have stomach pains, did you?"

"No."

"Let me have that," he said, and took my cup.

The second he walked away, I headed straight back to the bar.

After a year of doing open mike nights at Bonkerz, I started driving to clubs all around the state. Every big city had a comedy club with an open mike night, and if you got there early enough, you could get three minutes onstage. I managed to get four or five sets a week by doing Orlando, Cocoa Beach, Tampa, Miami, and Las Palmas. It was a lot of driving, a lot of time alone in the car, but I made the most of those hours in the car, listening to tapes of people whose voices I was doing.

While I was still at the radio station in Vero Beach, I got this idea from AA that if I made small, easily attainable improvements in the impressions I was learning, over time they would add up to major improvements. I thought of distinctions in the things that I learned, a tweak of how a vowel was pronounced, adding a nasal twang, moving a sound from the back of my throat to the front. I might make a single improvement a week, but that's fifty-two improvements at the end of a year. I never stopped thinking about it, and I never stopped trying. Tony Robbins talked about it as well. I never stopped listening to tapes. Between the ages of twenty-seven and thirty-nine, until the day I walked into my audition at 30 Rock, I drove all over the country, playing tapes, experimenting with my voice, with my tongue, with my teeth, with my diaphragm, with my breathing.

I also worked hard at writing decent jokes, but a lot of my stuff was cheap. Gay jokes. Fucking jokes. Time-of-the-month jokes. Dick jokes. Audiences would howl with laughter when I did the underhanded sales-man Mr. Haney from *Green Acres*, a 1960s sitcom starring Eddie Albert as a New York attorney who abandons city life to become a farmer and moves to the country with his glamorous wife, Lisa, played by Eva Gabor. Pat Buttram, who played Mr. Haney on the show, had a very distinctive, high-pitched nasal way of speaking, so he was instantly recognizable when I had him saying the phrase "Hershey High-way," which, in case you're not up on your Urban Dictionary definitions, is slang for the, um, back pas-sage. I was ashamed that I was putting that stuff out there, but it's all I had. Later I considered that kind of shit blackmail material. I would be walking down the hall at *SNL*, and I'd be afraid someone would come up and say, "I saw you one time in Daleville, Alabama, do a joke about Pat Buttram on *Green Acres* saying, 'Mr. Douglas, if I may introduce you to the Hershey Highway.' Was that you?"

I imagined responding in a haughty British accent, "I beg your pardon! You are mistaken, desperately mistaken. Good day, sir!"

I started getting booked in clubs around the Southeast. Sometimes the clubs were really sports bars where the game was on the giant TV, the bar was serving gigantic Long Island Iced Teas for 75 cents, and people were fucked up out of their minds, making out, screaming, and I'd be up on the tiny stage going, "So, any birthdays?"

People weren't even looking at me, but I'd have to stand there and pretend that they were. You have to finish your routine. When I needed to, I resorted to horrible shit just to make it through the night. I would do the filthiest, most vile material just to see if I could get a response from the audience. On one occasion, the NBA playoffs were on, and I remember thinking, There's a reason they don't have the playoffs on during *The Phantom of the Opera*. That's why waitresses don't come around for last call and give patrons their checks during the big eleven o'clock number.

So I'd do this joke: "How many people here read *Penthouse* magazine?"

No response.

"How about those letters to Forum? They're unbelievable, right?"

I'd start out using a cross between Bugs Bunny and a Joe Pesce kind of voice:

Dear Penthouse,

I never thought it would happen to me, but I was skydiving the other day from 1,500 feet when lo and behold I noticed that my skydiving partner was not a man but a woman. Imagine my surprise when she sidled over to me in midair, unzipped my fly, and engulfed my rock-hard cock. Although small for an adult male, a mere nine inches in length, she nevertheless hungrily lapped up my hot gurgling seed.

A woman with blond hair teased within an inch of its life caromed off the tables in back on her way to the bathroom.

I'd switch to a southern drawl:

Dear Penthouse,

I never thought it would happen to me, but I'm a schoolteacher at a local elementary school here. And the other day a student came up and asked if she could use my #2 lead pencil. I said yes. Imagine my surprise when she dropped to her knees, unzipped my fly, and engulfed my rock-hard cock. Although small for an adult male at a mere seventeen inches in length, she nevertheless hungrily lapped up my hot gurgling seed.

A table full of frat boys stood up and howled at the television when their team's star forward sank a three-pointer from half-court.

I brought it home with my best evangelical voice:

Dear Penthouse,

I'm a preacher in the Baptist Church here in Melbourne, Florida. I never thought it would happen to me, but the other day I was giving a sermon, and a parishioner came up to me behind the lectern, unzipped my fly, and engulfed my rock-hard cock. Although small for an adult male, a mere twenty-four inches in length, she nevertheless lapped up my hot gurgling seed.

Someone put a John Cougar Mellencamp song on the jukebox. And then I'd say to the crowd, "I'm sorry, that was over the line. I didn't mean to offend any churchgoers out there."

Let us pray. Dear Heavenly Father, we are gathered here at Giggles Sports Bar along Route 70 in Daleville, Alabama. It's been quite an evening so far. A lot of drink specials going on. I was onstage a little while ago, and lo and behold I noticed that one of the audience members was staring at my crotch. Imagine my surprise when she walked up to me onstage, unzipped my fly, and engulfed my rock-hard cock. Although small for an adult male at a mere fifty-seven inches in length, she nevertheless hungrily lapped up my hot gurgling seed.

The couple in back were still sucking face, a waitress was cleaning up a spill at table seven, and the blond weaved her way back from the bathroom, not one of them so much as registering my existence.

I sometimes hit the road with a kid named Billy Gardell, who now stars in the sitcom *Mike & Molly.* Sometimes the clubs didn't want to pay us our measly $20, so Billy, who was a formidably sized young man, would confront the managers to make sure we didn't get stiffed. He was only eighteen then, but he already wrote brilliant stuff. One time I gave him a joke I'd written about Beaver Cleaver, and in return he gave me a far better joke. This was back when the Ayatollah Khomeini had died: "I hope he's reincarnated as a tree, cut down, turned into paper, and on him is printed a copy of Salman Rushdie's *Satanic Verses*."

People would go crazy. They didn't notice the hot gurgling seed joke, but they went for the patriotic thing.

Bill and Pat Mullaly and I used to study the big names doing their stand-up routines to see how many laughs they got. With a stopwatch in hand, we watched Jerry Seinfeld's routine at the University of Florida and counted ten laughs a minute, and he'd *end* with an applause break. Jay Leno and Bill Cosby got that many too. It was incredible.

I won a stand-up contest for Funniest Comic in Orlando. My prize was being sent to the Laugh Factory in Los Angeles, where I would do a spot on *Comic Strip Live*, a popular stand-up showcase that aired weekly on Fox starting in 1990. The producer came to see me do my set. In the middle of it, I saw the silhouette of him throwing his arms up in disgust as he rose and left the room. He called my manager, who told me later he'd said, "How dare you send me this fucking amateur?"

One night at a club in Orlando where I'd done a whole bunch of impressions, an attractive young woman comic got onstage after me. She pulled out the largest carrot I've ever seen and said, "I'd like to see Darrell Hammond imitate *this*." The place went crazy. I had to agree, it was pretty fucking funny.

So I married her. A couple of times.

And then life dropped another bomb in my lap. One day in 1991, Dean, my sponsee, came to talk to me. He was in a bad way after having just been fired from his job. Apparently he'd been caught kissing another man, although why that was a firing offense, I have no idea. I guess it was that this wasn't the West Village, this was uptight Orlando, and Dean was a married guy with two kids. That night, he told me he was in love with me. I was shocked. When I told him I wasn't gay, that seemed to be the last straw. Dean used to joke that if he ever wanted to kill himself, he'd use a .357 Magnum. I never found that very funny, and less so when he finally did it. His wife was too distraught to give me the news. I can still hear his seven-year-old daughter's voice on my answering machine, telling me her daddy had killed himself.

I. Freaked. The. Fuck. Out.

In that moment, I thought I could hear flies buzzing in my ears. My senses seemed to cross boundaries with other senses, a smell I could feel, a fear I could hear. The whole room seemed to fill with the blare of a thousand strident trumpets.

I ran down to an AA meeting in Winter Park, but the meeting didn't work. I ran into Dean's sponsees, and I had to tell them what happened.

He had a closed-casket funeral.

I read a book called *In Tune with the Infinite*, which said that the first hour of the day, your mind is a clean sheet of paper, and you can write whatever you want on it. So I read positive-thinking literature like Emmet Fox's *Seven Day Mental Diet* for an hour every morning as soon as I woke up. I wouldn't allow my brain to think its own thoughts, or I thought I would lose my mind.

I did that for a year, but I no longer had faith in God or in the Program. I went to Overeaters Anonymous meetings too, because for a while after Dean's suicide, I couldn't eat, and then I couldn't stop. Eventually I quit going to all meetings, and I started to drink again. Five years of really good, productive sobriety, gone. My life had been better than I ever could have believed: I was making good money doing voices, I had a beautiful wife, I had a gorgeous apartment where I lay out by the pool slathered in suntan lotion, an item you would not find in my medicine cabinet today. I felt Dean's suicide was directed at me, and it worked. I had to leave Florida.

It seemed like a good time to give New York another try.

Seven years had passed since I'd flamed out of New York. I had a lot more going against me now, not least of all that I was in my mid-thirties, which is way too old to start a comedy career. And did I mention I was drinking again? It bears repeating.

My first gig in New York was at a short-lived club in Yonkers, a suburb of New York City.

The home of Mary J. Blige, Ella Fitzgerald, and DMX—and with W. C. Handy Place right off of Central Avenue—what were they thinking? Can you imagine poor DMX, a black Jehovah's Witness, living in Yonkers? I can just see him knocking on the door of some Irish or Italian family on Tuckahoe Road, trying to spread the word of God. His only hope was if he picked Steven Tyler's family home on Pembrook Drive. Like he was commenting on an *American Idol* contestant's performance, Tyler would have loved that: "I thought your preachin' was great, just great, DMX. You sang like a duck and you quacked like a singer, fiddle-de-dee kiss my little finger."

Anyway, the gig was at a place called Grandpa's Shooting Stars. Owned by Al Lewis, the actor best known for playing Grandpa on the sitcom *The Munsters*, although he did a fine turn in *Car 54, Where Are You?* before that, it was a classy joint with fishbowls on the tables, and the drinks all had long, long straws. The audience sat there sucking on these giant straws, and after eighteen minutes of a forty-five-minute set, they started to boo. I heard someone in the audience say, "He's wounded, finish him off." I slunk off the stage. At least I got carfare to get home.

And it was still the best gig I got. I was turned down by every club in Manhattan because I did impressions, and they were prejudiced against that. It was seen as a novelty act, and serious comedy clubs don't like novelty acts, guitar acts, prop acts. People looked down on it.

When I worked at the Skyline Motor Inn when I first lived in New York, I was living uptown in Washington Heights on 163rd Street. When I moved back to New York with my wife, we started out even farther uptown, at the very northern tip of Manhattan in Inwood. But eventually, after we realized living together wasn't working out that well, I ended up on my own in an apartment at 688 Tenth Avenue between Forty-eighth and Forty-ninth streets in Hell's Kitchen. Coincidentally, my building was across the street from the Skyline Motor Inn.

One afternoon when I didn't have much else to do, I was browsing through a bookstore, looking for something to read. Like my father, I was always drawn to stories of law and order, anything from Wyatt Earp and the frontier justice of the Old West to the sordid true crime tales of modern organized crime. I was trolling this section of the store when I saw a paperback with a photograph of the street sign on my corner on the cover. What were the chances? I had to get it.

The author was T. J. English, an Irish-American journalist who wrote for *Irish American Magazine* during the 1980s, and his book, *The Westies: Inside New York's Irish Mob*, chronicled the bloody misadventures of one of the most brutal gangs in the annals of New York organized crime. Unbeknownst to me, the Irish mob ruled Hell's Kitchen during the 1970s and '80s, exactly when was I working at the Skyline. According to the New York City Police Department and the FBI, the gang was thought to be responsible for as many as one hundred murders over a twenty-year period. Their big thing was chopping their victims up. Oh, and torture.

According to English, in the 1960s, a Hell's Kitchen native by the name of Mickey Spillane ran the organization. To be perfectly clear, this was not the beloved crime novel writer who invented the fictional hard-boiled detective Mike Hammer portrayed by Stacy Keach on TV during the 1980s. No, there wasn't anything make-believe about *this* Mickey Spillane, and he ran his crew much the way the Dapper Don, John Gotti, would run the Gambino crime family years later—while he operated gambling and loan sharking rackets, tossing in assaults and murders as needed, he also served as benefactor to the local community, earning his nickname "The Gentleman Gangster" by doling out turkeys on Thanksgiving, thereby ensuring the loyalty of everyone around him.

Things got interesting in Hell's Kitchen in the early 1970s, when Spillane's power was challenged by a young thug by the name of Jimmie Coonan, whose father Spillane had roughed up. Spillane fled to Queens, and Coonan took over. Spillane was eventually shot to death outside his apartment in Woodside moments after saying good night to his twelve-year-old son, Bobby, in 1977.

Coonan eventually joined forces with "Fat Tony" Salerno, who was a big cheese with the Genovese crime family, to split the proceeds in the construction of the Jacob Javits Convention Center at Forty-second Street.

Coonan later buddied up with the Gambinos, a partnership that was later taken down by an ambitious Italian-American federal prosecutor by the name of Rudolph Giuliani.

Coonan got sixty years for his trouble. So did one of his top enforcers, James McElroy, aka Jimmy Mac, the guy who didn't like to pay for his drinks at the Skyline. Jimmy Mac recently died in the California prison where he was twenty-five years into his bit. I read that at his May 2011 funeral at the Church of the Holy Cross on Ninth Avenue and Forty-second Street, a fair-sized crowd of people from the old neighborhood showed up to pay their respects to the man known for driving what they called the "meat wagon," which ferried body parts of the Westies' victims to be dumped on Wards Island, a lovely oasis in the middle of the East River known for its massive psychiatric hospitals that house the city's criminally insane, a sewage treatment plant, and, because Robert Moses may have been the most interesting city planner of all time, a lovely park.

And all this time, as far as I was concerned, these guys had been polite, tipped well, and encouraged me to do stand-up.

Even after I read the book, nothing I learned tarnished my feelings for those guys. Years later, one of my dearest friends was Bobby Spillane, that twelve-year-old kid whose father was murdered. Bobby, an actor who had appeared on *Law & Order* and *NYPD Blue*, kept me sober a hundred times over the years. In the summer of 2010, Bobby fell to his death out a window of his sixth-floor apartment on Fifty-third Street and Eighth Avenue. He'd just returned from a vacation, and his unpacked bags were still inside the apartment. At the time, Bobby had been collaborating with my old *SNL* office mate Colin Quinn on a one-man show called *A Hell's Kitchen Story.* The show died with him.

Just because I was drinking didn't mean I wanted to, so I kept fighting the sobriety fight. I attended a regular AA meeting on Forty-sixth Street between Broadway and Eighth Avenue. One of the guys who was also a regular there was supposedly a hit man, who claimed to have whacked eighteen people. He and another guy who was the brother of a well-known gangster had their own meetings in the back, because they didn't want to share in front of others. The hit man wanted to whack someone in the meeting because he didn't like him. He told this to Jackie, the big queen who ran the place.

"Look, I have the virus, so I'm not afraid of any of your shit," Jackie said. "You ain't whackin' nobody in my fuckin' meeting, you understand me?"

Everybody laughed.

"People are trying to get well, and you want to whack somebody? What the fuck is your problem?"

Anyway, the kid never got whacked.

Then there was Tommy, who used to brag that he was the best car thief on Long Island. He fell in love with a gorgeous Colombian girl named Valentina.

About six months after he started going out with her, I went up to him and said, "Tommy, you and Valentina are the best couple."

"Yeah, but Darrell . . ."

I said, "What? You guys are awesome."

He said, "You know, when I started going out with her, she wouldn't let me touch her down there. After a few months, she did. She's got a dick, Darrell."

Oh.

"By the time I got down there and found out, it was too fuckin' late. I already loved her!"

After meetings, a few of us would go to the diner on Forty-fourth Street and Ninth Avenue. Hell's Kitchen might have been rid of the Irish mob by then, but before Rudy Giuliani became the mayor the neighborhood had so many drug dealers and prostitutes and psychos wandering the streets, it was a seriously dangerous place to walk around. So my pal Sebastian, who was the queeniest drag queen you have ever seen, used to walk me home.

If someone said shit to us, Sebastian would snap his fingers and say, "Sweetie, I'm more of a woman than you'll ever have and more of a man than you'll ever be, so fuck off!"

Sebastian explained to me that a really successful pimp or a really successful drug dealer doesn't want to deal with a crazy person. It could fuck up the whole night, and they could lose thousands of dollars. Don't get into a fight with someone who's screaming at you, just let the crazy person pass. Made perfect sense to me.

Later, when the dealers found out I lived on Forty-eighth and Tenth and they started to know my face, they said, "No, no, he lives here. Let him through, let him through. He's neighborhood." They didn't want trouble, they just wanted to sell. I used to joke with people that I could buy anything on the planet on my stoop.

A week after Giuliani came into office, police vans started showing up, and these young buck cops came piling out, slapping all the drug dealers around, putting cuffs on. They made something like 130 arrests on my block alone in Giuliani's first three months.

I was told that the guys would get booked, somebody would pay their bail, then Giuliani would have them arrested again. Finally, they got tired of the game and moved on.

Hell's Kitchen became the best neighborhood in New York City. There is every kind of cuisine, every kind of perfume, every kind of music, every color of person, every religion. Once the neighborhood was cleaned up, residents returned to their stoops. It's a marvelous thing to see people in their lawn chairs having a bottle of beer with their grandkids playing nearby because it's safe.

The comedy career was not going the way I'd hoped. Then I heard about a booker working out of some dumpy office in Times Square next to a whorehouse who booked gigs in New Jersey. So I went to him, talked to him for a while, and he decided to give me a shot. I didn't even have to audition for him. Thanks to him, I made a living wage doing one-nighters on the west side of the Hudson.

And then I did a one-nighter with a dude named Dennis Regan, a fellow Floridian on the stand-up circuit who had previously owned an asphalt business with his dad called Tars and Stripes. Dennis's younger brother Brian was, and is, a great comedian as well.

Dennis told me, "I'm a regular at the Cellar. Do you work out there?"

"Yeah, at fucking three o'clock in the morning with non-English-speaking tourists facing the other way and drunk on schnapps or going through methadone withdrawal."

When people are blotto, or they've been dragged in off the street for drink specials, they don't give a shit about what you're doing onstage, even if it is a bona fide comedy club and not some Podunk sports bar. Woody Allen once said the audience has to know they're an audience. They have to come there specifically to be an audience, and know that they're expected to laugh. If they don't pay for that privilege, you're in trouble.

Dennis said, "I'll get you a decent audition. I think you're funny."

That was my first real break in comedy in New York. I got a spot at 9:00 p.m. instead of 3:00 a.m., and I did well, and I got booked into the Comedy Cellar on Macdougal Street in Greenwich Village thanks to Estee, the manager. Her full name is Estee Adoram, but every comic in the universe knows her just as Estee. She gave a lot of people their break. Even comics who have made it big go back to the Cellar to try out new material—Jerry Seinfeld, Chris Rock, Robin Williams,

Sarah Silverman. Seinfeld filmed a lot of his documentary *Comedian* there.

The cramped room with the iconic brick wall behind the stage is extremely well run; the audiences—the early ones, anyway—come because they really want to see comedy.

Rather than feature a single headliner, the Cellar is known for its showcase approach, booking several comedians for each show. Back then a single bill might include Ray Romano, Dave Attell, Dave Chappelle, Wanda Sykes, and, if I was lucky, me. I wasn't even doing impressions anymore, just straight stand-up. I only had one brief impression in my act: "I want to say to the people of America and the nations of the world," and then in Bill Clinton's gravelly drawl, "I hate you."

By the early 1990s, I started getting booked at Dangerfield's on the Upper East Side and the New York Comedy Club downtown in Murray Hill. Pretty soon I was making $75, $100 a night in cash. On weekends, the clubs paid about $50 a set; if I did six sets in three clubs, I could make $300 cash a night. If you worked hard, you could get by and pay your rent. That's what we were all doing.

It was exhilarating to be a New York comic, to be onstage with those guys, all of us just hitting our stride. I was very proud that I was on the same bill with them.

They were all really good people, but I didn't hang out with them much, since I tended to slink off after my set to nefarious dive bars so I could drink bad liquor and get blind. The bartenders at my regular hangouts all knew my address so they could tell the cabdrivers where to take me at the end of the night. Hell knows, there were a lot of times I was in no condition to summon up something as challenging as where I lived.

And then one night in 1995, Marci Klein saw a set I did at Caroline's near Times Square, and my whole world changed.

CHAPTER FIVE
It Was the Best of Times, It Was the Worst of Times

New York City
1996–2000

S ome time before the beginning of my second season on *SNL*, fate smacked me upside the head again. My wife and I were in the best-friend stage between our two marriages, which is how I came to be dating Alison, a twenty-seven-year-old blond Kathy Ireland lookalike with a penchant for sticking her finger down her throat. We had broken up because I was trying to do the tough-love thing they teach you in twelve-step programs.

"I can't stay around and watch you fill jars up with vomit and hide them during the night. I can only be with you if you get help."

The next day, she was swinging from a rope in her apartment on Fifty-seventh Street.

God might have been telling me, "Not so fast, Mr. Hammond." But the show must go on.

My second *SNL* season opened during the run-up to the 1996 presidential election, and much was made of the fact that President Clinton was well ahead of his Republican rival, elder statesman Bob Dole, in the polls. Ross Perot ran as an independent, as he had in 1992, but aside from giving Cheri Oteri material for *SNL*, his campaign didn't have much impact this time around. Clinton's enormous popularity was a huge boost for me professionally.

In the cold open of the first show, I appeared as a smug Clinton in an interview with Tom Hanks as Peter Jennings, in which Jennings declares Clinton the victor with "zero percent of the precincts reporting in." This was before the insane 2000 election, when so many news outlets incorrectly predicted Gore's victory, only to have to withdraw their comments in the wee hours of the night. But in the piece, Hanks/Jennings declares Clinton the winner two months before voters went to the polls.

It was the first of fifteen times I'd be putting on the white wig of the forty-second president that season alone, sometimes more than once in a night. I continued to suffer major lithium bloat, although it was still working for me as Clinton.

But it was more than my ability to impersonate people that contributed to my success on the show. I worked really fucking hard. I continued to live by the mandate I'd given myself at that radio station in Vero Beach all those years ago, that a small improvement every week would add up to major improvement in a year. My drinking sometimes got in the way, but I still managed to make progress. I was going up two or three hundred times a year at the Cellar to work on new impressions, and Caroline's had a 1:30 a.m. show on Saturday night, so after *SNL* I'd walked the three blocks across town and do a set there.

I also owe a lot to the people who helped me, who perhaps saw some talent or promise in me when other people thought I sucked. In October of my second season, *SNL* veteran Dana Carvey—a master impressionist who had famously done George Bush the First and the Church Lady on *SNL*—came back to host the show. The writers had come up with a sketch in which Dana played Johnny Carson, as he had a number of times during his run as a cast member, and I was to play Phil Donahue. In the sketch, Donahue goes to Carson's house to pick him up for a game of golf, but Carson refuses to leave until he finds his house keys. Dana was a bit of an angel, because the way it was written, Donahue was the stronger role. He could have had the sketch changed, or stopped it from going on the air if he wanted, but he didn't. He's a kind guy, and he understood that I needed help to solidify my place on the show.

It was a huge break for me. Remember, I was older than everybody, and I came from nowhere. A lot of cast members came from Second City, but I was essentially some guy Lorne found in a field on the plains in Somalia whom he'd seen kick a coconut, and he thought, That guy kicks pretty well. We should bring him back to New York, give him a job.

Again, if someone doesn't walk into your world and pull you out of the fucking mess that you're in, you could try forever and not make it in show business.

In December 1996, we debuted the *Jeopardy!* sketch in which stupid celebrities aggravate the hell out of Will Ferrell's beleaguered Alex Trebek. This recurring sketch would become a fantastic showcase for Will. Host Martin Short was Jerry Lewis, Norm Macdonald debuted his acerbic and filthy Burt Reynolds, and I did a blustery Sean Connery, although he was not yet the lascivious insult-flinger he would become in later episodes. The double entendres (and single entendres) were left to the Burt Reynolds character. Mostly, a combative Connery threatened Trebek like an eighteenth-century swashbuckler, and all three of us answered the clues so stupidly that at the end Trebek says that money would be taken away from our charities.

In May, we did *Jeopardy!* again. Will reprised his Trebek, Norm his Reynolds, and host John Goodman was Marlon Brando. The Connery character hadn't caught on yet, so I appeared as Phil Donahue, which had gone so well for me when I played him against Dana Carvey's Johnny Carson earlier that season. In the sketch, Donahue refuses to answer the questions, instead rambling on as though he were on his own talk show.

When we did *Jeopardy!* in the fall of my third season, in October 1997, Will reprised his role as Trebek and Norm his as Reynolds, but I ventured out as a dim-witted John Travolta, with *Friends* star Matthew Perry playing *Mr. Mom* Michael Keaton, who made faces by way of answering the questions.

It wasn't until our fourth outing with *Jeopardy!* at the end of that season, in May 1998, that Sean Connery returned. He would be there the next ten times we did the sketch, including my very last episode, when Will came back to host the show in 2009. Host David Duchovny played an airheaded Jeff Goldblum. Norm had left the show a couple of months earlier, so the third contestant would be a rotating crew. In this one, Molly Shannon had a go at Minnie Driver.

With Burt Reynolds out of the lineup, the writers let Connery be the sex-obsessed blowhard that he would become known as. As the sketch opened, Connery tellingly misread a psychology category: "I'll take 'The Rapists' for $200." In a later episode, he would misread "The Pen Is Mightier" as the "Penis Mightier." After Trebek corrects him, he says, "Will it really mighty my penis, man?" And even as Trebek further tries to set him right, he declares, "I'll order a dozen!" In later episodes, the Connery character would take it up a notch, never missing an opportunity to declare that he'd been banging Alex Trebek's mother.

I've often said that if you're not prepared to lose a finger, a thumb, or any other body part during an *SNL* workweek, you're probably not going to do well. There is tremendous drama all the way through to the very end, with the entire world watching. Models and political figures and world leaders and athletes and dogs and aardvarks and llamas and dancers, people in period costumes with parasols and high hats, a Model T Ford rolling by, Hillary Clinton walking by, Tom Cruise walking by, and then after a week of grueling work, you get cut from the show. You feel like your career is over, and you sit in your dressing room, trying not to cry.

That's what it was like during the third episode of my second season, when I did a couple of pretaped voice-overs but didn't appear on the show. It didn't happen that way a lot, but when it did, it was brutal. Later that season, I was slated to be in a fantastic cold open, but host Helen Hunt, who was hot off the success of her soon-to-be-Academy-Award-winning role in *As Good As It Gets* with Jack Nicholson, decided she wanted to do a bit singing Christmas carols instead, so my piece didn't make it. When I found out, I wanted to shoot myself. A few weeks later, I had a great piece with Sarah Michelle Gellar where I played Gene Shalit, the *Today Show* movie critic with the ginormous handlebar mustache, and she and I slow-danced, but it got cut. Being cut from a show feels like the first time a beautiful young woman calls you "sir." It takes getting used to.

You have to go through it for a while to realize Lorne didn't hire you for no reason. He knows what he's doing, and he wants you there. The writers are trying to get you in the show, even if they can't always do it. Maybe the set wouldn't come from Brooklyn in time. Maybe the host wanted to say "Live from New York." Maybe the musical guest couldn't act. There are a hundred reasons you might still be in your dressing room biting your cuticles instead of on the stage when 11:30 rolls around.

There were times when I felt good about the work that I'd done after the fact. I felt like a hitter going to bat with two outs in the ninth inning of the seventh game of the World Series every week. You don't get up in the morning doing joy jumps, thinking that's what's going to happen today. I rarely did character sketches except for occasional bit parts, I only did impressions, so that about cut in half my ability to get on the air. Some of the cast developed partnerships with certain writers—Will Ferrell had Adam McKay, Adam Sandler had Tim Herlihy. Given the special nature of what I was doing there, I didn't have that special kind of relationship with any of the writing staff. No one's bad, that's just how it was.

I might wake up Saturday morning at 9:00 a.m. and think, Oh my God, I just figured out how to do so and so. The smallest trifles in the world could improve a piece of four-minute theater. So I'd go in early and try to get those changes made. *How do I get something rewritten? Who do I go to? I'd like to add a line. I'd like to subtract a line. I have a new voice. I figured* *out a way to do the voice I couldn't do last night, but I can't get in touch with the writer.* All the sketches need to get to the cue cards around noon, and the cue card guys only do rewrites at certain times during the day, so I had to find out when *that* is.

Also, I was horrible at reading cue cards, which were there because the material changes so much up until air. And you have to play the cards but look like you're not playing the cards.

That's when I realized one of the reasons *SNL* is so difficult is because you're playing for two audiences in two vastly different places—the studio, and at home. My tendency was to make what I was doing play for the camera, which meant make it small, keep the gestures simple. A camera is so sensitive that you can think something, like, I'm annoyed, and it will show up on camera. With the camera there, it's very difficult to fake it.

On Thursdays and Fridays, all the writers gather in the writers' room to look at each script and suggest improvements. Everybody rewrites everybody else's material. Somehow, it was never contentious. Everybody behaves. You have to take it with good cheer when you get a shot to the teeth and your sketch is nixed. You have to keep trying. If you don't have that kind of personality, you can't be there.

On the wall outside the writers' room where we had our read-throughs are photos of cast members that, taken together, trace the history of the show. In addition to household names like Phil Hartman and Dana Carvey and Jane Curtin, there are lots and lots of talented people on that wall who couldn't deal with hitting a ball into the seats but not being allowed to run around the bases, which was what it felt like when a successful sketch didn't make it to air. It happened to everyone.

When a player is first on the show, the makeup department makes a mold of your face and head. If you walk down the *SNL* hall by the dressing rooms, there's a shelf with blank mannequin heads marked with current cast members' names, as well as those of frequently recurring hosts like Alec Baldwin and Tom Hanks. It felt like I spent more time in the makeup chair than other people, given all the impressions I did, although to be honest I never timed it and I have no idea if that's objectively true. There were some shows when I'd appear in four or five different sketches, which usually meant four or five different wigs and as many putty noses. I used to have the most painful rash blossom on the side of my face after shows because they'd rip a wig with, say, Sean Connery's sideburns off my head fast to get me ready for the next sketch as Rudolph Giuliani. My face got pretty burned up over the years.

It wasn't uncommon that I'd go home empty-handed Wednesday and not get handed an impression until much later in the week. I'd call in to see what was going on, but it could change three times between Friday and Saturday. Back then, when I wasn't sure I was going to make it, those were torturous days.

From the minute I got the phone call that I got *SNL*, I knew I was living on another planet for the rest of my life. This was a different world, a world where dreams come true and every club you play now, they're paying attention. Since it's too physically demanding to do more than two or three live *SNL* shows in a row, there are a lot of dark weeks during the season. For me, that was time to hit the road with my stand-up act. And I wasn't relegated to the bowling alleys and honky-tonks and road-houses anymore. Now it was the Punch Line in Atlanta, Cobb's San Francisco, Zany's Chicago, Improv Chicago. I played some splendid clubs all around the country.

I also got called to appear on *Letterman, Leno, Conan,* shows that had rejected me just a few years earlier, before I got *SNL*. And when I finally did debut on *Letterman,* I did the same set that I'd been doing all long. When I was first invited to sit down with Dave after my set—which is the equivalent of knighthood for a stand-up comic—he asked me to do Donahue. Before I did that appearance, knowing that I'd probably be invited to the couch, Norm Macdonald said, "If you get jammed up out there, just say anything. Dave will turn it into a laugh." So I tried it.

"This is easier than I thought it'd be."

Dave said, "Speak for yourself."

Big laugh.

"You didn't have to interview Cybill Shepherd earlier."

Bigger laugh.

Applause break.

Cut to commercial.

Genius.

I suppose I shouldn't have been surprised when I received an invitation to Clinton's second inauguration in January 1997, although the White House might well have rescinded it if they'd seen the sketch I did two days before, in which Clinton stood in a pantsless police lineup for Paula Jones and yelled "Live from New York!" with a paper bag on his head.

Going to Washington for the inauguration was pretty heady stuff. I flew down on the shuttle with Katie Couric, who was then cohost of *The Today Show*, and other big shots from NBC and was booked into the first five-star hotel I'd ever been in, the Jefferson, which is only a few blocks from the White House.

I walked into my room feeling humbled and on top of the world at the same time. And then I saw a display of Grand Marnier.

This is how fast my brain did it: *I don't think I've ever had Grand Marnier.*

Click.

Next thing I knew, I was watching Katie do an interview at the night-before party when someone came up to me and said, "You're going home."

The next thing I knew after that, I was lying on the floor of National Airport in my tux, and I could not stand up. People kept coming up to me and asking, "Are you okay?"

Each time, I said, "Yeah."

Finally I asked someone to help me up, and I guess somebody did. I don't know if it was the same person or someone else who walked me to my gate. When I got there, I couldn't find my ticket.

I'd had diamond studs in my tux, and they were gone.

Apparently sensing my distress, a woman, who I can only imagine recognized me, came up to me and said, "Are you going to New York?"

I somehow indicated that I was and that I didn't have a ticket.

"I'll buy you a ticket."

She sat next to me on the plane back to New York, and gave me money for a cab. If I'd been more coherent, I'd have gotten her name and number so I could thank her properly when I sobered up. But if I'd been more coherent, I wouldn't have needed her help.

By midnight I was back in Hell's Kitchen, drinking a quart of Budweiser. The next morning, the forty-second president took the oath of office, and I wasn't there.

So it wasn't just in my professional life that other people kept saving me. It seems like my life was filled with angels who walked into my own personal purgatory and conducted me safely out of there. It's how I ended up living this long.

I was still going to meetings, trying to be sober, but after the two suicides of people I loved, it was really hard. I'd stay sober for months at a time, and then I'd get drunk. I wasn't a guy that drank for days on end. I usually got drunk as fuck one night, and then I'd go back to meetings and I'd stay sober for a few more months.

It was surely thanks to whomever it was that got me out of the inaugural festivities before I caused any damage that I subsequently received an invitation that

April to appear at the White House Correspondents' Dinner. I should clarify that it wasn't Darrell Hammond, the person, who was invited—it was Darrell Hammond as Bill Clinton.

The dinner is a Washington institution, where the White House press corps gets to hang out with the subject of their daily attention in a collegial setting. The president always makes a funny speech, and there's usually a comedian who pokes fun at all the luminaries on the dais and in the audience. That year, Norm Macdonald had that honor.

I was called in to play Clinton's clone because a few days before the event Clinton had fallen downstairs at golfing great Greg "The Shark" Norman's house in Florida. Clinton had to undergo surgery on his knee, and he was still on crutches. Before the dinner, I went to the White House to meet him in the Oval Office—in full Clinton drag, the white wig, the fake nose, the works.

As I shook his hand, I could feel the flush of shame on my face. "I have to admit, Mr. President, I feel foolish."

He smiled that magnetic smile of his and said, "I think you look *terrific.*"

The dinner starts with various awards given out to journalists, and then the president makes his speech. A few minutes into Clinton's, he complained that his leg hurt, so he called on his "clone" to take over at the podium.

"Bill, will you come out here?"

As I walked out from backstage, with "Hail to the Chief" playing me on, I thought, There are too many powerful people here. Anything could go wrong. This ain't like blowing a set at a bowling alley in Daleville, Alabama.

The White House wrote my material, which alleviated my jangly nerves, if only a little, when I stepped up to the microphone. "Mr. President," I said in Clinton's voice, "have a seat."

The first time I ever performed an impression in front of the person I was imitating, and it had to be the most powerful human on the planet? Really, God? Couldn't we have started with someone a little less, I don't know, able to have me deported?

In the end, the worst thing that happened was CSPAN misspelling my name "Darryl" in the broadcast that aired later.

I was invited back as the featured entertainment for Clinton's eighth and final White House Correspondents' Dinner in 2000. That was way scarier than my previous appearance. For starters, I wasn't in character this time, I was just me. And I had to come up with my own material. By then the nation had endured the Monica

Lewinsky scandal, the graphic details delineated in the Starr Report, and the president's impeachment—the first time that had happened since Andrew Johnson in 1868. (Nixon probably would have been the second if he hadn't jumped ship before the House got a chance to vote.) For a comedian, that kind of stuff is gold, but it's a different beast when you're standing next to your victim and he has the power to have your taxes audited back to the Stone Age.

The previous year, in the spring of 1999, Ms. Lewinsky had made a cameo appearance on *SNL* in a sketch with me. In it, the soon-to-be-ex-president dreams about his postpresidential life as a playboy in Malibu, tethered to Monica, the new Mrs. Clinton. And yet here I sat on the dais with the honored guests of the evening, about five seats away from the actual president.

(Incidentally, my mother called me after the Lewinsky episode to say, "You didn't really kiss her, did you?" Yeah, Ma, we all know where her mouth has been. Thanks for reminding me.)

Clinton himself broke the ice when *he* had a go at *me.* "Poor Darrell," he said. "What's he gonna do when I'm out of office?"

There was one joke I told that I worried about. I was talking about what it would be like to have Clinton's incredible magnetism as he says to a woman, "If you would only take your clothes off and let me see you naked, there would be no more racism." In my mind, there was a deadly silence in the room as everyone waited to see the commander in chief's reaction. In truth, he guffawed louder than anyone, and gave Linda Scott, an African American news producer seated to his left, a big hug for added effect.

As I said during the show, the president of the United States laughing at my jokes—amazing.

Despite my outward success, I was still struggling to understand the sense of torment I felt. The treatment I'd been under, both therapeutic and pharmaceutical, wasn't helping, and neither was the drinking, so I went to a new psychiatrist.

He said, "I want to talk to you about multiplicity," doctor-speak for multiple personality disorder. "I want to talk to you about losing periods of time."

Who said I was losing periods of time? I never said I did. No one's ever come up to me and said, "Hey, remember the day you were in the Santa suit, and we went to Yankee Stadium and then we flew to Alaska?" Or, "I saw you on the corner of Twenty-seventh and Third yesterday," and I have no recollection. Sure, I blacked out sometimes when I was drinking, and yes, I'm good at assuming different voices, but what am I now, Sybil?

CHAPTER SIX

God, If You're Not Up There, I'm F*cked

The Bahamas
1993

There have been a few times in my life where I felt I was touched by God. The first was in high school after the condom broke while I was having sex with Miriam Bonaparte on the beach in Melbourne. Instead of immediate terror, I felt this sense of calm, that it was okay. (And it was.) The second time happened when I was walking to work in Orlando when I was twenty-nine, and I felt God saying, This is all going to work out. The third time was in 1993 when I heard a Bahamian judge say, "Will the defendant please rise?"

In the early 1990s, my manager got me some gigs for cruise lines. I started with Carnival, then I went to Norwegian, then I did a few for Disney. The money was great, and the talent was treated very well. I don't think I paid for anything on the ship, and most of the time I was left to my own devices, since I only had to do two thirty-minute shows in a week.

The average age of the passengers is about 103. They've been herded around like baby geese all day. They're sunburned. They're ornery. They're wondering if the trip was worth the goddamned expense. Since most of my material was filthy then, it was tricky cobbling together clean jokes, so I did impressions, which is perfect. It's ten o'clock at night, liquor's on the house, they'll laugh at just about anything—as long as the material is broad, derivative, and shitty.

I didn't think anything of it when this old guy in the audience fell asleep during my show. I'd seen plenty of people in audiences catching a nap, and these were old folks with more rich food and liquor in their systems than they were probably accustomed to. I once did *Macbeth* in summer stock, the straw-hat circuit in Rhode Island, and people would routinely snooze during the show. The actors weren't bums, they were talented people. The guy playing Macbeth would be up there doing his most stentorian baritone, "Tomorrow, and tomorrow, and tomorrow . . . It is a tale told by an idiot." And you'd hear snoring from the seats. So when the geezer went face-first into his scampi, I was just grateful that his snores weren't drowning me out.

It wasn't until my set was done and the lights came up and the old dude didn't move that one of the waitresses went over and tapped him on the shoulder to wake him up. When she didn't get a response, she looked over at me and said, "I guess you killed."

Uh, that's not what I had in mind.

The ship's bar might as well have been the crew's quarters. A lot of those people had given themselves to the sea years ago. They were career ship guys. There were comics and croupiers and staffers, all kinds of folks who got lured into the idea of room and board for five years and good money and health insurance, and never left. I could see why. The ocean is hypnotic. But the isolation of being out there working as a comic and not knowing anyone on the whole ship can cause a person to, shall we say, compensate.

On one cruise, I was introduced to a cocktail called the B52. I don't even know what's in a B52—Kahlua, ouzo, cream, Clorox, rifle cleaner—but it'll set you right. The night I had my first B52, I joined the band for a rousing rending of "Honky Tonk Women," in which I did a fair-to-middling Mick Jagger.

There were a lot of English people working on the ships, and I was enchanted by them. I loved their manners, the way they spoke, especially the ones whose accents leaned a little toward cockney or North London. That accent was deliriously wonderful. I'd hear them refer to their mums, and think, I'm actually experiencing another culture for the first time. I used to want to go up to them and say, "Can *everyone* in your country act? Like *all* of them?" But I never did. Instead I drank B52s.

After my second or third B52, I danced, and I don't dance. There's a reason for that. One day when I was maybe thirteen or fourteen, I was hanging out in my room listening to the radio when some great song came on, Smokey Robinson or James Brown or the Supremes, and I started moving to the groove. My father suddenly loomed large in the doorway and glared at me as if I were responsible for the Black Death. I never danced after that. (A few years ago, I was invited to be a contestant on *Dancing with the Stars*. I called my agent and said, "Tell them I'll be happy to do the show as long as there's no dancing.")

But there I was out on the briny deep doing a fair Mick Jagger and dancing with some very interesting British people. Amazing what circumstance and sixty-seven B52s will do.

At that time, I used to get down on my knees and pray before I went out with the English girls to drink. *I know I'm going to get drunk tonight, but please—please—don't let me break my nose again.* I'd broken it the first time playing baseball in junior college, and then twice more on the cruise ships: once falling downstairs, and another time falling into a doorknob. There's nothing like being at sea with a broken nose.

One night, after another Rubbermaid garbage can full of B52s, one of the English girls gave me a copy of the book *The Shining*. Another girl gave me a tape of the sound track to *The Omen*. They said, "We like to smoke dope and listen to this. It's really freaky when you look out at the ocean, and you see all the black water." They thought it was a joke, but for me, it was a little too close to home. *Whoa.*

On my last Disney cruise to the Bahamas, the ship pulled into Freeport on a Thursday night, with nowhere to be until an hour before setting sail the next morning. Was it a coincidence that Freeport didn't exist before I was born in 1955? As far as Freeport is concerned, I wouldn't mind if the world wanted to go back to 1954. Who builds a major city out of nothing but fifty thousand acres of scrubland and swamp? What is this, Florida? I guess it makes sense, since Florida is only sixty-five miles away. Three-quarters of the Bahamian population lives in Freeport, meaning the laid-back emptiness of the islands you see in the travel brochures has been replaced in the capital by streets both narrow and heaving. But if you run a business there, the name Freeport is apt at least: no one's paying any taxes on anything till 2054, so it has the feeling of the Wild West, only with bigger ships, seedier bars, and a mighty chill accent.

On this particular Thursday, I found myself at a bar on the main drag. The place was filled with tourists, refugees from other cruise ships, locals. The bartender served these killer drinks made with *four* shots of rum apiece. In the space of a few delicious hours, I'd had four of them. I was having a blast.

I stumbled up a long, narrow flight of stairs to the bathroom to take a leak. A guy I'd been chatting with at the bar stumbled up the stairs with me. I didn't think much of it until he tried to sell me some coke. I didn't really feel like it, so I declined. But while I'm at the urinal with my dick in my hand, he's giving me the hard sell. "Just try a sample. You like it, tell your friends."

I didn't have a good feeling about it, but I was pretty wasted, and I really wanted this guy to go away so I could piss in peace. With the hand that wasn't holding my johnson, I reached into my pocket for the small wad of bills I still had and pressed them into the guy's hand. The guy stuffed the wad into his pocket and handed me a bunched-up dollar bill. I stashed the dollar in my pocket, zipped up, and headed back to the bar, relieved to have gotten rid of him.

I never made it down the stairs. Two cops were waiting for me outside the bathroom door. They patted me down, found the bunched-up dollar in my pants, saw that it was dusted with white powder, and put me in cuffs.

I was so drunk it seemed funny at the time. But it didn't take long for me to realize that the situation was no joke. A security guard for the ship watched me being taken away in chains like I was disappearing into a pool of sharks.

I don't remember much about the first jail, but in the morning I was transferred to another holding cell. It was about twenty by thirty feet and window-less, so I couldn't really tell if it was day or night. It was so hot it was hard to breathe. There were no beds, no chairs, no toilets. I asked the guard on the first day if I could go to the bathroom. When he took me there, I said, "There's no toilet paper." He screamed at me, "Do your business!" I thought,

Fuck me. Sixteen shots of rum—career move, Darrell. Nice going.

We had to sleep on the stone floor, which hadn't been cleaned since viruses were invented. I was sweating and filthy, and they'd taken my lithium away from me. I used my Reeboks as a pillow and finally fell into an exhausted sleep. I began to dream of the stream that ran behind our house in Melbourne when I was a kid, swinging out over the sun-dappled water on a rope with my friends, and the laughter and the color and the smell of the leaves and the trees. The utter joy. When I came to, I was still in that putrid cell.

Other prisoners came and went. At one point, one of the other prisoners said or did something to piss off one of the guards, and the guard hit him with a stick. The guy was bloodied and stunned, and he sat there for a long moment before he reached into his mouth and pulled out what looked like a tooth. Then he shit himself.

The captain of the guards came and whispered to me, "You're not in America anymore. You're going to do serious time."

There was a guy in the cell with me who'd driven his car into a store and killed somebody. If I'd been that guy, yeah, all right, you got me. But I'm locked up for partying?

I asked for a Bible. The harshest guard in the world is probably going to give you a Bible no matter how much of a scumbag they think you are. I started reading the Twenty-third Psalm. I figured it was worth a shot.

I'd heard one of the guards say something about my going before a judge on Monday, so I told myself if I could just get to Monday, I'd be okay. I walked back and forth in the dark saying, "One second less. One second less. One second less." I wanted it to end. In fear for my sanity and my life, I wasn't thinking about my root chakras, my 401(k), New Year's resolutions, or combination skin and frizzies. I didn't think I was going to make it. I'd rather they just shot me than leave me in the dark for one more day with other people's shit on the floor. They had put guys in solitary in Alcatraz, the toughest motherfuckers in the world, and they had broken in the dark and silence. I was just trying to not cry or yell. That was my prayer. *Hey, God, don't know if you're there, hoping that you are. If you are, I would like to not start screaming at my trial. If you aren't, I'm fucked.*

At some point they let me use the phone. In my desperation, I called my father, and as soon as I told him what had happened, he slipped into military mode, making plans to get me out, whatever it took.

The second night I was in jail, three big dudes came into my cell. They introduced themselves as being from some drug enforcement agency.

"DEA?" I asked.

"No, no. Not DEA." Without fanfare or really any interest in me at all, they said, "If you can give us thirty-five hundred dollars American cash, you can walk out of here tonight."

"I don't know anyone who has thirty-five hundred." I had just made $2,000 working on the ship, but I couldn't get to it. "My father is coming down here Monday. Can you wait till then?"

"No." And they left.

On Sunday night, one of the guards, an enormous man named Stanley, came to my cell and in this sleepy voice called my name.

"What are you doing here?" he asked.

"I got popped," I said. "I had a dollar bill that they said had coke in it. I don't know what it was."

"Man, you're not a criminal. You don't belong here."

Stanley took me out of the cell and gave me barbecued chicken and collard greens. Up until that I had only had grits and ham and a cup of water three times a day. The chicken was glorious. And he gave me a *USA Today* sports page.

"You know you've got about a fifty-fifty chance of making it out of here, right?" he said.

No, I didn't know.

"No prior offenses anywhere?"

"No."

"If you get the right judge, he'll fine you. Wrong judge? You will stay in that cell you're in two to three years."

Then he turned and started talking to his friends. I couldn't really follow what they were saying, but they seemed to be religious, God-fearing men. I heard one of Stanley's friends say, "Make me an instrument of Thy peace."

I finished my meal, and Stanley said it was time to go back into my cell. "Good luck tomorrow."

I'd been in there three days without my meds, and my mental state was deteriorating fast. Stanley's act of kindness got me through that final night.

Monday morning I was led across a dusty courtyard to the courthouse in chains while everyone else was going on about their lives. Talking. Laughing. Slapping one another on the back. Making plans. Eating. Smoking. And I wanted to say, "Excuse me, hello? There's a man in chains here?"

When I arrived at the courthouse, my father was standing there with a briefcase filled with cash.

I looked at him and said, "This is gonna be okay, right?"

"I wish I could tell you that," he said, "but I don't know." It was really not the answer I was hoping for.

While we were talking, a guy came up to us and said in a rich Bahamian accent, "I'll get your son off for five hundred dollars."

My father didn't miss a beat. "I'll give you two hundred right now. And I'll give you another three hundred if you get him off. That's your five. You can get him off?"

The guy said, "Yes, I can." Very confident.

I thought, It's that easy to get a person out of this near-death experience?

I had no choice but to trust him.

I must have inherited something from my father, that I walked into that courtroom with my head up.

In the courtroom, my "lawyer" argued that I had such a small amount of whatever it was, it couldn't even be weighed. "It's just dust, your worshipfulness," he said.

The judge seemed to consider the paperwork in front of him with practiced expertise. He looked down at his desk, and then across the room at me. He said, "Will the defendant please rise?"

I stood, and as I did, I felt that familiar sense of calm, God's presence or whatever it was. I guess it's what you feel when you nod at the hangman and say, "Okay, I'm ready." You are at that moment of acceptance.

The judge fined me $1,500 and told me that as soon as I paid and signed a few papers, I'd be free to go.

The guy who sold me the drugs was in the prisoners' docket with me. The bailiff, who must have been new, said, "What about him, your worshipfulness?"

The police and court officers laughed, and the judge waved him off like he was an idiot for asking. "Charges dropped!"

As my father and I were leaving, we saw the dealer smoking a cigarette with the cops on the courthouse steps. My father looked at him with an expression that said, *You know I'm killing you, right?*

On the plane back to Florida, my father worked out his revenge plans, figuring out which guys he'd known from the war were still alive, guys who for the right price would function as mercenaries.

"I'm going to rent a house down there, and we're all going to go in there, and everyone who works in that jail is getting killed. We're going to make sure that fucking captain is there. He's getting killed. And that guy that gave you the fucking drugs, he's getting killed. I just have to work it out."

He was planning to smuggle in an arsenal a little bit at a time by boat, get an apartment to stash everything in, and then go down to that police station and kill everyone. I wondered briefly if he'd actually do it.

To me, that was just Dad. That's how he thought.

The night I got back to New York, I got a good drunk on at Dangerfield's. Drinking rum for only the second time in my life (because that first time had gone so well), I got it into my head to head downtown to the Ravenite Social Club—words to chill the hearts of any law-abiding mobster caught bad-mouthing the Gotti family at a Bensonhurst barbeque. Located at 247 Mulberry Street in what is now the last sliver of Little Italy that hasn't been swallowed by Chinatown, the club was where John Gotti held court, sort of the Gambino crime family's HQ. That is, until the killjoys at the Federal Bureau of Investigation bugged a third-floor apartment, and bingo—down came the indictments. It didn't help that underboss Sammy "The Bull" Gravano turned state's evidence, either.

My favorite story I ever heard about the Ravenite is the bomb someone left outside the club for Gotti in 1989. "A gift for John Gotti," the thoughtful card accompanying it read. The "bomb" turned out to be a box of salt, basically, wired without a full circuit. That said, you have to hate someone pretty hard to leave a bomb outside their club, especially when it's the head of the Gambinos and he knows how to dismember you. (I guess some people are still pissed off. There was a comment posted in May 2011 on YouTube under a video about the Ravenite that reads, "I'm gonna cut your balls off and make you eat 'em you fuck, R.I.P JOHN GOTTI THE KING OF NEW YORK.")

But I was fascinated by guys who didn't take shit—boxers, soldiers, cops, Mafiosi—so I was thinking, *Maybe the Don will let me work for him. And after a while, he'll whack those guys for me.* That's how much rum I drank.

If you ask a cop, he's going to tell you that people who drink too much can think crazy thoughts. I once left a bar in Mexico after a night of drinking that tequila with the worm in it, because I was convinced I had to get up early to try out for the Yankees. The same thing happened to me years later when I got a colonoscopy, and the doctor gave me the same stuff Michael Jackson was taking when he died: propofol. It turned out to be one of the best things that ever happened to me.

I went to my shrink afterward and she said, "Look at you! What's going on with you? You seem so clear and bright and crisp."

I was writing jokes in the cab, thinking, I'm going to write the greatest comic novel of all time, or I'm going to learn to pole-vault—today. Either one. I think I'm going to pencil that in. Going down the West Side Highway, the cab passed the Trapeze School in Hudson River Park at Houston Street, and I thought, How come I never did that? Well, there's still time. I could totally do the Flying Wallenda thing.

I was noticing the architecture. Wow, I've got to read that Frank Lloyd Wright someday. Yeah, I'm going to do that. Read Frank Lloyd Wright, learn how to be a trapeze artist, write a monologue for Jay, and pole-vault.

I even forgave the nurse for asking me in her heavy Slavic accent as I was going under with a camera about to be stuck up my ass: "What would Clinton say now?"

Anyway, the last thing I remember about that night in Little Italy was getting out of the cab and seeing four imposing gentleman in suits standing by the door to the Ravenite Club. I woke up on a sand pile in the alley behind the club. I looked myself over to see what kind of shape I was in. My wallet was still in my pocket, the money in the billfold. But then I looked down and saw that my sports jacket was covered with dark red splotches. As I sat up in alarm, the situation became clear—those red splotches weren't blood, they were marinara sauce. Evidently, I'd eaten well.

When my daughter was born, it wasn't so much a mystical touched-by-God moment as it was empirical proof of His existence. When I saw my daughter's face for the first time, that word came to me: *impossible.* But I was starting to understand that my whole life was about impossible shit. In any given week, I faced the impossible. *I've got to write an eight-minute set for my appearance on* Conan. *I have to learn four impressions on SNL. I'm trying not to drink or drug. I'm having nightmares every night. Some nights I wake up screaming. I change my sheets every day because I sweat so much.*

If I'd ever had any doubts about God's existence before, that changed when I saw my daughter being born. More to the point, it changed when I saw a little head appear in my wife's pelvis. Women talk during this procedure, about centimeters and contractions and minutes apart.

I wanted to say, "Goddamn, honey, you're doing math while a head is emerging from your vagina?"

It's very, very impressive.

And if the baby can't come out, they cut you, and not on the knee. *In a very strategic area.* If a man were cut in the same area, there would be homicide in the delivery room. Possibly a suicide. Nobody would get out of there alive. I remember trying to figure out the amount of pain I was looking at, and I finally decided it was the equivalent of getting hit in the mouth by the space shuttle.

I said to the nurse, "Why don't you give her some Demerol?"

The nurse looked at me and said, "Demerol can't handle that."

"What are you talking about? In the army, a guy gets shot, they give him Demerol."

"That's right," she said, "but around the maternity ward, we call gunshot wounds pretend pain." The look on her face said, *So why don't you sit in the corner, eat your potato chips, and shut the fuck up?*

Why don't women brag about having a child? We men, we fix a doorknob, drinks are on the house. We really think we're going to turn on the television and hear Ted Koppel say, "Tonight, against all odds, Darrell Hammond, using only a six-piece ratchet set, let himself back into the house that he also locked himself out of."

When I got home from the hospital, I sat down and wrote on a piece of paper:

Schizophrenic
Manic-depressive

Borderline personality disorder
Major depressive disorder
Multiple personality

I taped the list on the back of the door and sat on the floor with a bottle of Jack Daniel's just looking at it for a while. These are the things I had been diagnosed with; these are things I was being treated for. How do I overcome all these things so that I can be a father?

My wife and I were so excited for this baby. Although we were divorced at the time, when I found out my wife was pregnant, I knew immediately I wanted to be that baby's daddy. It was like every electron, every neuron, in my body went YES! Hell, I even married her again before the baby was due.

And when at last I held that child in my hands—and there's a photo of me, horribly bloated, dissolute, holding this perfect infant girl—my world rocked. How could something so perfect end up like me? If this were possible, then the world could not be as I thought it was. The pristine nature of this creature was not possible. But mess that I was, I was going try to take care of this baby, even though I had no idea how to do that.

I didn't realize that ninety percent of what you have to do as a parent is either trial and error, or in your genes. It's in your hands and your cells. It's thousands or millions of years old. The first time my daughter cried, I panicked. But then something inside me said, *She's a baby. She wants you to be in charge. Pick the baby up, pat the baby on the back. She wants to be comforted.* How did I come up with that? Well, it's either in my DNA or it's common sense. I picked my daughter up, and I comforted her. I didn't put a knife on her tongue or hit her in the stomach with a hammer.

Wait, I didn't *what*?

CHAPTER SEVEN

Blood on the Floor

New York City
1998

After my daughter was born, I started experiencing flashbacks, often at work. For most of my years at *SNL*, I had a windowless corner office just off the writers' room on the seventeenth floor of 30 Rock. When a flashback struck, my mind would instantaneously travel back to the kitchen of the house on Wisteria Drive, while the office floor and walls seemed to turn red, red and wet—like blood.

My home wasn't any safer. Sometimes I woke up during the night, and I didn't recognize my bedroom or my clothes in the closet. I looked at my feet and had never seen them before. Other times, I would wake in the dark and I could feel invisible hands pushing down on the mattress, or I'd see the silhouette of someone standing just inside my bedroom door. When I turned the light on, no one was there.

I was haunted by horrible nightmares. The cold sweats I'd been having before only got worse. I was changing my sheets so often I started to keep a fresh set beside the bed. For the sake of my family, my wife and daughter didn't live with me but in an apartment nearby. I couldn't subject them to the sound of my screaming in my sleep.

I kept a pint of Rémy in my desk at work. I never drank right before or during a show, but sometimes, when disjointed childhood memories popped into my head, I sought refuge in the bottle. I drank on a Saturday morning, had breakfast, then slept for a while before going to rehearsal. The drinking calmed my nerves and quieted the disturbing images that sprang into my head.

When the drinking didn't work, I cut myself. The wound created a fresh crisis to get me out of the one in my head. It gave me something else to focus on. It was more blood, but it wasn't *that* blood.

I had a careful system for cutting: I prepared by laying out a razor blade, two square gauze pads, and a couple of strips of precut surgical tape (I learned the hard way that the tape gets twisted if you try to cut it with your teeth while trying to stanch an open wound). I used Good News disposable razors because the plastic top was easy to break off so the blade fell out. It was perfect for a quick slice and easy to control—I could cut only a little, or I could cut a lot. Over the years, I accumulated dozens of scars, narrow and neat, or raw and jagged, like angry ladders across my arms, legs, and chest.

Usually I made small cuts, but when I was drunk, the cutting could get dangerously deep and I couldn't stop the blood flow. I once approached Gena Rositano, the stage manager, when I needed help with a bad cut, but usually when that happened I'd take myself to the NBC infirmary. One time I had cut myself really bad, and I still couldn't get out of the flashback. I told the nurse, "I feel like I'm little. Everything's red. Something terrible happened." I didn't know who I was talking to or where I was.

She did not seem suitably impressed, although I don't know what I expected her to do. I started screaming like, well, a madman.

Next thing I know, there are two cops there, hands on their guns, and a paramedic taking my blood pressure before strapping me into a straitjacket. My wife came, but I didn't recognize her until we were in the ambulance on the way to the hospital.

I should have gotten myself an E-ZPass to the ER at New York Hospital, I was there so much. It's hard to tell one visit from another. I can say that once I calmed down, I was released and went back to work. It might have been a day or two, but that's all.

The staff there got to know me. I'd walk in and be like, "Jake! What's going on?"

I remember on one visit, when I wasn't feeling quite so jolly, a nurse asking me, "How are you?"

How did she think I was? I was tempted to point to the sign over my head—"Emergency Room"—which seemed explanation enough.

Usually a staff member would go through my bag to make sure I didn't have any drugs or weapons. A psychiatrist would examine me, but I was always able to convince the doctor that everything was okay, even though it clearly wasn't. No one knew what to do with me. Every doctor I saw essentially shrugged and wrote out another prescription for antidepressants or antipsychotic medication.

"Darrell. Darrell! DARRELL!"

I was standing on Broadway, my infant daughter in my arms, and my wife was shaking me by the shoulders.

"The baby's crying," I said, as though I heard the thundering hoofbeats of the Four Horsemen of the

Apocalypse. My wife had gone to the drugstore, and the baby started crying, so I had picked her up and left the apartment to find my wife, only to walk right past her on the street. I didn't know who she was.

"Babies cry, Darrell. It's not a big deal," she said, taking my daughter from my arms and heading back to the apartment like nothing was wrong. And nothing was. I have many times thanked my lucky stars that my daughter has a good mom.

The sound of my infant daughter crying almost always triggered a flashback. My brain must have made an emergency connection to the sound of a child in distress and linked it to the demons I carried from my own childhood. The result was that I often overreacted. She might merely have been hungry or in a need of a diaper change, but I would be filled with the kind of adrenaline that a soldier needs to charge over a hill with a gun and bayonet.

On another occasion, my wife and I took our daughter to visit some friends in Orlando, and the baby began to cry in the night. I ran into their kitchen and grabbed a knife. I was certain that something was coming to get us.

My wife stopped me and said, "There's no one here. It's just us."

It can't have been fun to live with somebody whose brain was under siege.

Then one night while my daughter was still an infant, I got very drunk, and I sliced my arm open far worse than I usually did. Strangely, it turned out to be one of the best things I ever did. In the ER, a psychiatrist I'd never seen before came to examine me. After listening to my itemized list of psychological ailments for about thirty minutes, she said, "You're not a lithium candidate. You're not manic-depressive. You're not schizophrenic. You're not psychotic. You do not have borderline personality disorder."

What?

She explained it all in a way no one had before.

There are three kinds of bipolar disorder. Type I means you've had at least one fully manic episode with periods of major unhappiness. This is the one they call manic depression. Type II is when you've never had full-fledged mania (and for that I'm sorry for you, because the highs are fucking high). Instead you've been pretty high, just not super high. And you get blue, too. The third type is something called cyclothymia, a mild form of bipolar that makes it burn when you pee and gives you tenderness in the testicular regions. You get it from unprotected sex with an infected person; a dose of antibiotics, and you're as right as rain.

Schizophrenics are exactly what you think they are: they're the real deal. They're not a bit blue, or a bit wacky—this is a nasty illness, where poor bastards can't tell what's real and what's not, have fucked-up reactions to just about everything, and can't exist in a normal social situation. It's not pretty, and I'm glad it wasn't me.

Psychosis is just a fancy way of saying the patient doesn't really know what's going on—it can be part of bipolarism or schizoid behavior—it's like the bloom on the flower of mental instability. Probably the worst part of psychosis is the hallucinations. You want spiders crawling up the walls? Sign up for psychosis. Again, thank you, God, for keeping me away from that shit.

Borderline personality disorder? You think in total extremes at all times day and night twenty-four/seven forever. Please don't abandon me; I fucking hate you; wow, you're the best; and then there's the cutting, and the overdoses, and all that. It was the one closest to some of the things I was feeling, but still it didn't seem right. Though I was "turbulent," I was also often better than turbulent.

Finally, she said, "What do you think you would have to do to someone so that they couldn't speak or hear? You're a trauma patient. What brought you to this ER again was something that happened to you."

We will be able to hew out of the mountain of despair a stone of hope, as MLK said.

For the first time in my life, the problem was going to be treated—the *real* problem. By this time, I had been to twenty-three shrinks, and none of them had been able to make the correct diagnosis; they had all simply gone back to their prescription pads and given me another kind of medication.

After I was released from the hospital, the psychiatrist began to see me as an outpatient. But in order to treat me, she said, I had to revisit the trauma that had made me sick. We started going back through my life. She knew what she was looking for as if there were a bullet lodged in me, and she needed to find it to identify the gun and the shooter. People who come from healthy homes don't open up their veins.

That's when the memories started to come back.

I am barely past toddlerhood, and I wander out to sit on the wooden ties of the railroad tracks behind our house. I practice crying while hoping one of those eighty-car freight trains will run me over, flattening and distorting me like a penny left on the track. My mom stands by the door of our house, forty yards away, and watches me with a quizzical expression, as though she is curious to see what will happen. I suppose she's close enough to grab me if she hears a train coming, but I don't know.

Sometimes I find myself standing in a swamp filled with water moccasins about a mile from our house. I have no memory of how I got there.

Each night, my mother stands in the doorway to my bedroom and recites the poem "Little Orphant Annie," written by James Whitcomb Riley in the 1880s. The poem was inspired by a young girl named Mary Alice Smith, whose father was killed during the Civil War. Riley's own father was serving in the conflict when his mother heard about young Mary and invited her to come live with the Riley family in Indiana when James was a little boy. As was the custom, the little orphan "Allie" did chores around the homestead to earn her keep.

Thirty years later, Riley's poem was published in the *Indianapolis Journal* under the title "The Elf Child." It began:

> Little Orphan Annie came to our house to stay,
> And wash the cups and saucers up,
> And brush the crumbs away,
> And shoo the chickens off the porch,

And dust the hearth, an' sweep,
And make the fire,
And bake the bread,
And earn her board-and-keep;

And all us other children,
When the supper things are done,
We sit around the kitchen fire
And have the best of fun
Listening to the scary tales
That Annie tells about,
And the Goblins will get you,
If you don't watch out.

Once there was a little boy
Who wouldn't say his prayers
And when he went to bed at night,
All the way upstairs,
His Mommy heard him holler,
And his Daddy heard him bawl,
And when they turned the covers down,
He wasn't there at all!

And they searched for him in the attic,
And the cubby-hole, and press,
And they searched up the chimney,

And everywhere, I guess;
But all they ever found
Was his pants and round about
And the Goblins will get you, if you don't
    watch out.

Once there was a little girl
Who liked to laugh and grin,
And make fun of everyone,
Her family and kin;
Whenever there was company,
And guests were sitting there,
She mocked them and she shocked them,
And said she didn't care!

Suddenly she kicked her heels,
And turned to run and hide;
There were two great big Black Things
Standing by her side,
They snatched her through the ceiling
Before she knew they were about!
And the Goblins will get you, if you don't watch out!

Somehow these terrifying stories of what happens to children who misbehave was transformed in the 1920s into the comic strip *Little Orphan Annie* and the Raggedy Ann doll. Later the stories were adapted again into the sunny Broadway musical *Annie,* in which the orphan girl cheerfully sings "Tomorrow" to her little dog Sandy. The show also became known for launching the career of a fourteen-year-old Sarah Jessica "Sex and the City" Parker.

In my mother's version, the little boy and girl aren't snatched up by goblins, they are butchered. All that remains are bits of blood and hair.

I wake up in the middle of the night to find my mother leaning against my bedroom door, staring at me. She says nothing.

We are at the beach, and a boy is being carried out to sea. The boy's mother is screaming hysterically, and people on the shore are shouting for help. There is no lifeguard in our area, but a man from farther down the beach hears the ruckus and sprints over. He dives in, swims out to the kid, and grabs him. He swims back with one arm, the other one holding the boy's head out of the water. As everybody on the beach cheers, I look over at my mom, who is silently watching the scene play out as though she were waiting for a traffic light to change.

My mother speaks on the phone with one of her friends in this lovely lilting southern accent, and

I think, She doesn't talk like that. When she speaks to me, her voice gets deeper, and the accent disappears.

My mother walks into the living room nude. She knows I am home. "What are you doing here?" she demands, her face trembling with rage. "I didn't think you were in here."

She says the same the thing the next time. And the next.

I am getting into the car to go somewhere with my mother. "Wait," she says, holding the door open. "Put your hand there." When I do, she slams the door.

My hand touches the electrical cord for the brown lamp in the living room. The cord has been cut, and I receive a fierce shock when my fingers meet the exposed wires inside.

I am three or four years old, and my mother is holding me close to her with one arm. In her free hand she holds a serrated steak knife. Slowly, she sticks it into the center of my tongue, making an incision about one-quarter inch to one-half inch long. It is quiet except for the sound of the hibiscus bush thump-thumping against the kitchen window. I do not struggle or cry.

Somehow I know that to do so will make it worse. The kitchen floor is red with my blood.

I am four or five years old, and I'm in a park in Jacksonville, alone. When I get home, I am bleeding from my penis and my ears, and I have a 104-degree temperature. My abdomen is swollen and purplish. I will not let my mom or dad touch me. My father goes to get Myrtise so that I can be put into a tub of ice. Myrtise cleans the blood off and puts me to bed. My mom tells everyone that I had a reaction to penicillin.

My mother removes the safety plastic that covers the electrical outlets and sticks my fingers in the holes.

When I am five, my mother hits me in the stomach with a hammer. She tells the doctors I hurt myself playing.

"Do you want to hit me?" I ask a kid on the first day of first grade.

"What?"

"I'd like to make friends with you, so do you want to hit me?"

"No."

I don't understand his answer. I think that's how you get people to be nice to you, you let them hit you first.

The year I turn ten, my mother becomes enchanted by the title of a noir thriller called *I Saw What You Did* starring Joan Crawford and directed by William Castle, who would later produce *Rosemary's Baby*. In the movie, a couple of teenage girls have nothing better to do than make prank calls. Whenever someone answers, they whisper, "I saw what you did, and I know who you are." Then they reach a man who has just murdered his wife; panicked, he decides to find the girls and shut them up.

Sitting on the couch doing homework or watching TV, I glance up to find my mother standing there looking at me, and without her southern accent, she says, "I saw what you did, and I know who you are."

Through my work with my new therapist, I started to understand why I'd grown up surrounded by a sense of evil, firm in the conviction that I was going to be killed. I woke up every morning in fear for my life. I went to school every day in fear for my life. Went to bed at night in fear for my life. I knew I couldn't say anything about what was happening. Who would believe me? My mother believed in Jesus. I was certain they would tell me, "You're a bad boy."

My mom discovered that if you want to control someone, you have to make them beholden to you. Relieve them of their fear, relieve them of their pain. In my case, she had to give me the fear first, give me the pain. She made sure that I believed I was the problem, that I was a bad person, and that's why these things were happening.

In *The Grapes of Wrath*, John Steinbeck describes a scene that hit home when I first read it. Al Joad is taken aback when he sees his big brother, Tom, for the first time in four years when he's paroled after killing a man with a shovel. Tom Joad now wears the expression of a prisoner who doesn't give anything away: "the smooth hard face trained to indicate nothing to a prison guard, neither resistance nor slavishness." I learned to exist in my house as if I weren't there at all. I kept my voice a level monotone, my expression flat. I didn't reveal anger or fear or sadness. If I showed any of those emotions, it suggested that something was wrong in that house, and that made my mother angry.

After a while, I started to think that if I could make myself cry, she'd lose interest in making me cry herself. If she thought I was already hurt, she might leave me alone. I practiced crying sitting out on those train tracks behind our house.

What finally worked were the voices. I could win her approval, or at least avoid a beating, by doing this one thing she liked to do. Bob Cratchit and Ebenezer Scrooge spared me her rage. When I demonstrated that genetically we were wired to copy sounds, her eyes would go soft and childlike as I said my line. She would listen to me intently like a mechanic listening to an engine.

I learned how to do voices from her—they were my only protection.

As my treatment continued, dredging up these memories, the flashbacks kept coming, like quick and violent snapshots of the past. Sometimes I would see images as two-dimensional drawings. My doctors have told me that sometimes POWs, trauma survivors, incest survivors, remember Picassoesque versions of things.

In Barry Levinson's 1985 film *Young Sherlock Holmes*, the teenage investigator looks into a string of deaths seemingly caused by hallucinations. In one scene where Holmes and Watson are tripping, the king of hearts gets up and walks out of the deck of cards on bendy cartoon legs, which freaked me out because I had had visions of my mother exactly like that.

One night, a friend knocked on my apartment door. I grabbed the lamp beside the bed, ripping the cord out of the wall, and went looking for the sound. Something in my soul was saying, *It's not going to happen to me again.* I didn't even know what "it" was, but the sense of looming danger was unmistakable.

I was in my apartment one night, stabbing myself in the leg, when one of my pals from the Hell's Kitchen AA meeting, Big Mike Canosa, knocked on my door. He was a big tough guy who ran around with the Westies.

"Gimme the knife, Darrell."

He told me that staying sober, trying to do the right thing by your family, showing up for work, was far more difficult than drinking half a bottle of tequila and shooting some guy in the back of the head. He said the longshoremen in the neighborhood who were still getting up and going to work every morning and abiding by the law and taking their children to church every Sunday, *they* were the tough guys.

When my daughter was less than a year old, my wife and I took her to Tampa, where I was doing a corporate job—a private performance for employees or clients of whatever company had hired me. I was doing as many corporate events as I could in my off weeks to bank money for my kid's future. My mother and a group of her women friends came to this one. The ladies passed the baby around the table, each one cooing at her in turn, the baby giggling and gurgling with happiness. When my mother picked her up, my daughter's face went blank. No laughing, no smiling, nothing. It was like she knew.

Eventually my wife and I tried to live together in a house upstate in Tuxedo Park, New York, a small village on the National Register of Historic Places. Situated in Orange County in the Ramapo Mountains, it is about forty miles north of Zabar's on the Upper West Side of Manhattan. The ubiquitous penguin suit that men are required to wear at formal functions is named after the place, as it was here that the first sucker, a guy named Griswold Lorillard, had the audacity to wear a black-tie outfit that didn't have tails to the Tuxedo Club's annual ball.

We had a gorgeous Tudor-style home in a secure gated community, two greyhounds my wife had rescued from a track, our beautiful daughter, and for a few years it was lovely. I remember sitting in our leafy yard, reading the Sunday papers, nibbling on these gigantic muffins Elaine Stritch had sent by way of thanking me for getting her tickets to *SNL*, while my daughter played with neighborhood kids down the street.

As with all good things in my alcoholic world, our upstate idyll didn't last. I started staying in the city during workweeks, driving up on weekends after the show. And eventually my wife missed the twenty-four-hour amenities of the city, so she wanted to bail too.

I started reading books on cognitive therapy that talked about self-fulfilling prophecies, and I began to see that my vision of the world was distorted. It wasn't true that only horrible things happened, and that death was always just around the corner. That's the way I had lived my whole life. Now I realized I felt bad merely because of the way I was thinking.

Tony Robbins said that you should ask yourself what you would have to believe to feel bad about your situation. When I got up each morning, I would write down my disempowering thoughts, the ones that had totally overwhelmed every happiness and success that I'd achieved. Then I'd write down all the positives: I had a fantastic career, a beautiful daughter, a nice home, plenty of money. I began to believe that it was possible for me to have a good day.

Joel Osteen, who is a very gifted speaker who draws his advice from the Bible, said, "Respect your parents." And I thought, Until they stab you.

After several years of trauma therapy and going back through this forest of dark memories, my doctors thought it was important for me to confront my parents with what I was learning. For days I agonized about calling them. How do you bring this up with the people who raised you, the people who presented themselves, publicly at least, as your biggest supporters, the people who allegedly cared more about you than anyone else on the planet?

When I finally summoned up the nerve to dial the number that had been etched in my memory since I was old enough to count to ten, my mother's response was simple and heartfelt: "Don't ever call us again." Her accent was notably absent.

I would not see them again until my mother was dying and my father wasn't far behind.

CHAPTER EIGHT

What You Didn't See

New York City
2000s

M inutes before *SNL* went on air the night of
October 7, 2000, I was in my dressing room
putting on the expensive suit I was to wear in the
cold open when *FLASH*, the floor turned red. I was
playing Vice President Al Gore in the first presidential debate against George W. Bush (Will Ferrell).
Suddenly, a year of studying and practicing was down
the drain—I couldn't remember how to do the voice. I
couldn't remember what Gore sounded like or what I'd
seen in any of the tapes I'd been studying ever since it
became apparent a year earlier that Gore was going to
be a front runner in the millennial election. I couldn't
remember what he looked like or why I even *wanted* to
do the voice.

I had to do something.

I got out the gauze pads and tape, and popped the top of a razor.

I made a thin cut across my left forearm. It wasn't horribly messy, and I wrapped my arm fairly carefully, because I knew I had to wear that nice suit for the sketch. (At *SNL*, whatever the person you're impersonating wears, you wear, no cheap imitations.) The point was to buy me enough focus so that I could go out there and do what I wanted to do with that character. Luckily Jim Downey, who wrote the sketch, had been giving me line readings during the week, so I knew what I had to do by muscle memory if nothing else.

In the sketch, I ramble on and on in Gore's peculiar southern drone about the "lockbox" he planned to put Medicare and Social Security in. He had talked about this metaphor for protecting the country's senior citizens during the actual debate earlier in the week, and Downey elaborated on the idea to make it seem more literal—with keys to the lockbox, and an explanation of where the keys would be hidden, including one under the bumper of the Senate majority leader's car. I had also devised an awkward stiffness in my movements in caricature of Gore's real posture.

Will ended the debate with the word that Bush felt would describe a Bush presidency: *strategery.* The word became so entrenched in the public psyche during the real Bush's presidency that people thought he'd really said it.

I said, "Lockbox." A reporter for *U.S. News & World Report* said I lost the election for Gore. I felt terrible.

Will and I received an International Radio and Television Foundation Award in 2001, which honors outstanding communications professionals for their accomplishments. There was a big banquet at the Waldorf Astoria Hotel, and the award was presented by Tim Russert. Past winners of that have included Diane Sawyer, Katie Couric, Charlie Rose, and David Letterman. And now us?

Let me be clear: I don't say this to brag, I say this to point out how seriously some people take clowns. I mean, we made fun of George W. Bush and Al Gore, that's it. Perhaps the pundits would have felt differently about me if they'd known what this "expert" was doing to prepare for the sketch that supposedly did Gore in.

In the middle of all this hilarity, my old acting roots started gnawing at me, and I started thinking about returning, at least part-time, to serious drama. I fortuitously ran into Valerie Harper at NBC, and she told me about an acting coach named Harold Guskin, whom she had worked with. I went to him, and he had me read a monologue from *Death of a Salesman*. Partway through, he interrupted me.

"You have to hear it before you say it, don't you?"

I nodded.

"That kills it. You're obligating yourself to a sound, and obligation is the death of art. I'm going to work with you on *what* you're saying, not how you say it."

He gave me a line to read.

"I ran up eleven flights of stairs."

"How many?"

"Eleven."

"How many?"

"Eleven!"

"How many?"

"ELEVEN!"

"That's it!" he said.

The following week, I auditioned for an episode of *Law & Order: SVU*, and I got it.

I have always been careful to write jokes that will make everybody in the room laugh, Republicans and Democrats both. As a result, everyone thought I was on their side. That's why the newly installed Bush administration invited me to perform at W's first White House Correspondents' Dinner in April 2001, even though I had twice appeared at the same event during Clinton's presidency.

For me, the main difference was that at this dinner, I was even more nervous than I'd been in the past, so I had strategically consumed a bottle of wine beforehand. I suppose that's why I thought my performance went so smashingly well. As I told Chris Matthews, who graciously came up to me after the event to compliment me (I'd done him during my routine), it's unnatural to have that many powerful people in one room. At *SNL*, there are only 300 people in the studio audience, but at the Washington Hilton, there are 2,600 of the nation's most important players in politics and the media.

A few minutes into my act, I gave the new president a baseball glove and invited him to play catch, which he seemed to enjoy. He even forgave me for all the Gore material I did that night. I reprised some of the lockbox sketch, which went over very well on the Republican dais, so I turned to Mr. Bush and asked, "Can I do some more of him? I worked a year to learn the damn guy, and then you beat him." I was supposed to read some telegrams, but I had so many pieces of paper on the podium—oh, and that bottle of wine before—that I couldn't find them. I guess it really was a good performance, though. At the end, the audience gave me a standing ovation. Or maybe they were just getting up to go to the bathroom.

A month later, the *SNL* team decided to put together a Mother's Day special. The other cast members invited their mothers to appear on air, but I, obviously, did not. The writers thought it would be fun to have me play my mom instead.

Using photographs my sister supplied, the hair and makeup folks got to work. I'll never forget sitting at the mirror when the makeup artist began to paint my mother's face on my own before dress. When the hair person lowered the wig on to my scalp, I could hear my mother saying, "I saw what you did, and I know who you are."

I started vomiting right there in the chair. I was rushed to the infirmary, where my blood pressure was recorded as dangerously high. It was as though my body had said, "It would be preferable to die than to let this continue."

The show went on without me.

By the time 9/11 happened, I was barely functioning. But the entire city was in a state of grief and bewilderment; we were the same then. So many had died. A great chunk of downtown had been removed from the skyline by those planes—literally, in the space of two hours, removed, gone, disappeared, the people, the buildings, everything—and those who remained were scared out of our wits.

The season premiere of what would be the show's twenty-seventh—and my seventh—season was less than three weeks away. Nobody in the cast or crew, let alone the rest of New York, knew how to behave. How do you put on a comedy show when there's so much destruction and fear and sadness in the air?

Lorne invited Mayor Giuliani, who was in the process of becoming "America's Mayor" for the deft way he handled the crisis, and a bunch of NYPD and NYFD, still in uniform and covered in dust from Ground Zero, to be onstage in front of an enormous American flag, while Paul Simon sang "The Boxer" in tribute to New York City.

At the end of the song, Lorne turned to Mayor Giuliani, thanked him for coming to the show, and asked, "Can we be funny?"

Giuliani said, "Why start now?"

It was a stroke of genius. It allowed New York and the rest of the country to laugh again.

On the Friday night before our third show of the 2001–2 season, NBC News announced that anthrax had been found in *NBC Nightly News* anchor Tom Brokaw's office and in the post office in the basement of 30 Rock. It wasn't the first incidence of the deadly white powder turning up around the country— a guy had died in Florida already, and the *National Enquirer*'s offices had been shuttered because of it—so people knew exactly how to respond when they heard: they left the building.

Host Drew Barrymore, whose new movie *Riding in Cars with Boys* was about to be released, had already debated whether to fly in from Los Angeles for the show (she was hardly alone in the country in being afraid to get on a plane then). The anthrax was almost more than she could bear, and she vacated the premises too.

But me, I felt strangely calm. I guess I had become somewhat inured to the presence of danger, to the sense that something bad would probably happen. I already understood the notion that the danger you're guarding against is already inside. No need to lock the door, it's already in. And that was more true than we realized. By the time we found out about it, the anthrax had already been in the building for a couple of weeks. Apparently, some assistant to an assistant opened the envelope addressed to Brokaw, thought nothing of it, brushed the powder into a wastebasket, and went about her business, but not before she passed the letter to the real assistant to deal with. That assistant had it on her desk long enough for Brokaw to see it and joke, "If he thinks he's going to threaten my life, at least he could be grammatical."

Once it was determined that the danger was limited to the parts of the building where the anthrax had been discovered, Drew and everybody else got it together and returned to the studio. In her monologue that Saturday night, Drew thanked the studio audience for being brave enough to come to the building when it was all over the news that there was deadly powder drifting around. She also thanked her then-husband, Tom Green, for supporting her—and the camera cut to him sitting in the front row, wearing an enormous gas mask.

The show went off without a hitch, and, as *SNL* has always done when there's a big news story, the writers simply incorporated the anthrax scare into the show. I did the cold open as Dick Cheney. Ever since 9/11, the real Cheney had been in hiding, and there was endless speculation about where he was. Neighbors of the Naval Observatory, the vice president's official residence, complained about noise that suggested they were perhaps constructing a secret bunker there. So he might have been there. It was later said that one of the compounds where he hid is called the Raven Rock Mountain Complex, Site R, near the Pennsylvania/ Maryland border, a few miles north of Camp David. Apparently when you drive by, you might see the two enormous metal doors in the side of Raven Rock Mountain, like something out of an Austin Powers movie. On top of the mountain there's a ton of communication antennae and towers and stuff, making it easy enough to find—they might have thought of that. Inside, the compound even has its own water source—some huge lake or something. They built the whole thing during the Cold War in case the Russians nuked Washington—at least a few top officials could survive while the flesh burned off of the rest of us.

But since we didn't know any of that at the time, everybody was guessing. In the sketch, Cheney "reveals" that he's really been hiding in a cave in Afghanistan, personally hunting Osama bin Laden, because "old Uncle Dick is gonna make sure you don't have to worry about opening your mail come Christmas."

A few days later I went to Washington, D.C., where I was scheduled to play one college in the afternoon and another at night. That afternoon I got up in front of all these people, and here I was, this guy who made fun of politicians. No one wanted to make fun of the people who stood between us and the next attack, let alone hear someone else do it, and here I was, my whole act based on exactly that. My act was invalid, but it hadn't occurred to me before I went on. The afternoon set went horribly. Nobody laughed.

During my gig that night, I resorted to doing random impressions of Porky Pig and Barney Fife, the hapless deputy sheriff played by Don Knotts in the 1960s sitcom *The Andy Griffith Show*. These were bits I had performed more than a decade earlier when I was starting out.

But that night also saw the birth of a new impression: the gloriously unpredictable New York City gadfly, Al Sharpton. A Baptist minister and civil rights activist, Rev. Sharpton is as famous for his track suits and bouffant hairdo, styled after his hero James Brown (whom he did a killer impression of on *SNL*), as he is for his demonstrations whenever any incident even suggests a racial motivation. Sharpton had drawn attention to a number of racially charged incidents, from the murder of three African American men in the Howard Beach section of Queens in the 1980s to the shooting death of Sean Bell in a hail of fifty NYPD bullets on the eve of Bell's wedding in 2007. He's staged unsuccessful but always colorful campaigns for public office numerous times, from mayor of New York City to the U.S. Senate and even president of the United States. He never came even close to succeeding, but his speeches and debate appearances were always fun to watch. He was pretty powerful up there sometimes.

I'd been thinking about doing him for a while, and then I saw him on television in the days after 9/11, taking a run at Mayor Giuliani, who was at the time universally praised for how he handled himself in the wake of the terrorist attacks. On one of those Sunday-morning political shows, I saw Senator Dick Gephardt talk about Homeland Security. On another channel, I saw Senator Joe Lieberman talk about the Patriot Act. Then I turned on *Hannity & Colmes*, and I heard Al Sharpton say in his distinctive preacher voice, "I did not call Giuliani a bozo."

When I did him that night for the college, I pretty much just repeated what Sharpton had said verbatim, and the audience roared.

Eventually I would be able to work some of our national nightmare into my stand-up routine. I remembered how drunken audiences seemed to really go for a little patriotic bloodletting, so I came up with: "If we ever capture Osama bin Laden, instead of turning him over to the United Nations, we should kidnap him and drop him along a dusty road in Daleville, Alabama.

Instead of getting justice from the United Nations, he gets it from some farmer with a twisted leer and a misshapen head. 'You're a loooooong way from Tora Bora, ain't ya, boy? Jethro, get the Slip 'n' Slide! We're gonna love you!' " It was cheap, the crowds used to really go for it, but I'm not at all sorry that the joke is now obsolete, and not just because the dude is dead.

During the first week of December 2001, the city was still reeling. The smell of the fires still burning at Ground Zero permeated lower Manhattan, and remnants of nearly three thousand bodies were still being recovered.

That week, I went to the funeral of an old drinking buddy from Hell's Kitchen who had died at thirty-three of cirrhosis. For all the devastation I'd caused in my own life with my drinking, it was a blow to see the corpse of such a young man who had literally been killed by booze.

After the ceremony, I joined my friend Caryn Zucker, who was then–NBC head Jeff Zucker's wife, and *Today Show* host Katie Couric for dinner. I wanted so desperately to function at Katie's level, because her rarified world was supposed to be mine too, but I felt like she could see right through me. I was sweating profusely throughout the meal. I was fumbling to make myself understood. It was as though there was a line right down the middle of the table, Katie and Caryn sitting on one side, in the light, and me on the other, sitting in darkness.

After dinner, I went back to 30 Rock for a late rehearsal. Rock god Mick Jagger was the musical guest that week, and the writers were trying to work him into a couple of sketches. I had played Karl Lagerfeld opposite Maya Rudolph as designer Donatella Versace at the beginning of the season, and there was another sketch with those two characters slated for this week. But when Jagger came up to me and asked if I'd mind if he did Lagerfeld, I was both flattered and relieved. *Sure, have a go!* Normally, if someone tried to poach a part from me, I'd be pissed, but it was Mick Fucking Jagger.

Besides, I was still in a daze from seeing my friend in his open casket, knowing full well it could have been me.

CHAPTER NINE

You Want Me to Go *Where*?

New York City
2002

I got along great with everyone at *SNL*, but I didn't hang out with cast or crew at the infamous postshow after-parties because my partying wasn't really about partying. It was about getting obliterated, and I didn't want those people to see it. The flip side was that when I was trying to stay sober, it was hard to be around revelers because I couldn't join in. The alcoholic's catch-22.

But I was at work a lot, so when I was using—I'd started adding an obscene amount of cocaine to my binges—I had to be creative about how I did it without other people catching on or letting it interfere with the work. At least too much.

In those days there were beautiful young women who would walk up and down Sixth Avenue, visiting various office buildings to take orders for coke. Addicts themselves, they provided this service for the dealers in exchange for drugs. And the dealers they worked for supplied the purest, strongest stuff you could get. Once you started, you couldn't stop.

One night on Fifty-eighth Street, I met up with this tiny Russian woman, Irina, who'd been selling to me for a while. There was blood running out of her nose and over her lips from all the powder she'd snuffed up her own nose. She was in a kind of stupor.

"I didn't get all of your coke," she said. "But I got most of it. Can I blow you to make up the difference?"

At another time in life, perhaps, when confronted with a hot Russian chick saying, Hey, how 'bout if I blow you?—I couldn't imagine a context in which that would not be appealing. But then again, I couldn't have imagined such a girl having blood all over her face.

"Thanks, but you're bleeding, and I don't want blood *down there*. I like nice things *down there*. Like strudel."

She just looked at me dumbly, continuing to bleed.

"Why don't we get you cleaned up?" I said.

We went to a deli on Seventh Avenue and got some napkins. Then we went to her apartment, and we stopped the bleeding. While we were there, her phone rang—it was her supplier with the rest of the coke.

So he came over, and we did coke together. And then *my* nose started bleeding. For a moment, I stepped out of myself, like a tourist in my own life, and saw this pathetic scene: these three people in this squalid apartment ingesting lines of white poison in the name of a good time. I thought, I don't like this anymore.

My coke use got so bad that when I called Irina for more a couple of months later, she said, "I'm not selling you any more. You are out of control. You're famous, and I can't have that. I can't have some terrible thing happening and I was the last person you saw that day. I don't need it."

It's not a good sign when your dealer cuts you off.

I went to Las Vegas to perform at a Microsoft event a short time thereafter. I'd done a lot of coke and drunk about a fifth of vodka on the plane. There was supposed to be a rehearsal the morning I arrived, but I was in no shape to attend. I couldn't find the theater on my own, so someone was sent to fetch me from my hotel. When I was finally where I needed to be, I was unable to walk or speak very well. If they had only let me sleep for a few hours, I could have walked out there and behaved as if nothing had happened. They moved the rehearsal back to the evening to give me time to sober up.

But instead of going back to my room, I seem to have played the slot machines, because I ended up with a bucket filled with quarters. I tried to give my coin stash away to a group of Asian women, but I guess I didn't make as good an impression as I thought I did, my clothes rumpled, unable to keep my balance, slurring, and my hair standing straight up, a feat I would have been unable to accomplish with even the largest tub of styling gel.

Somebody from Microsoft found me and got me back to my room. I slept, and I came back that night as if nothing had occurred. I had a great rehearsal, we did the show the next day, and I got a standing ovation.

But waiting in the wings of the stage where I did my set was a man with airplane tickets to Minnesota, one for him and one for me. Apparently, the Microsoft people had called someone at NBC, who called my manager. As a result, the people in my life—my manager, my agent, my wife, and Lorne, I think—became concerned, and there was a bit of an intervention. The man with the tickets relayed a message: You have a great career ahead of you, but only if you take these tickets and go to Hazelden.

Founded in the 1940s by a recovering alcoholic named Austin Ripley, Hazelden was initially intended for the treatment of alcoholic priests(!), although that plan was quickly scrapped and replaced with the credo: "A sanatorium for curable alcoholics of the professional class." Fortunately, they're pretty relaxed about that definition, or they'd have stopped me at the door, as I'm anything but professional and I'm not sure you'd say I have much class. It might not be quite as famous as the luxurious Betty Ford Center in California, but Hazelden is one of the premier addiction treatment facilities in the country.

It was clearly an offer I couldn't refuse. I went.

I cursed myself that I didn't get drunk on the plane. Detox facilities actually prefer it if you show up under the influence—makes you easier to manage—but I didn't give a shit about that. I was just sorry I'd blown my last chance to drink. Instead, I spent the two-hour flight from Vegas to Minneapolis trying to grasp the concept of never drinking again. I knew I'd stopped for long periods before, but somehow this seemed more final.

When I got to Hazelden, the escort left me on the patio by myself. I sat down on the ground and started crying. There were other people in their first hours of detox standing around smoking cigarettes. No one said, "Hey, are you all right?" They were out of their minds. We were all out of our fucking minds.

After I went through an intake assessment, in which they determined what kind of addict I was and what my mental health issues were (it was a long meeting), they gave me Librium, which is given to help ease the agitation of withdrawal.

Men and women are housed separately. In my dormitory, there was one phone and between thirty and fifty men. I was initially supposed to have a roommate, but they ended up giving me my own room because I was still waking up screaming. (Also, I snore.) There was another guy there who had his own room. He had been a drug dealer who'd found a pregnant woman who had hung herself and the baby had fallen out of her. It freaked him out so badly he could never sleep at night.

As part of my treatment, they put me in a trauma group with people who had been raped and tortured. I had to go every day and listen to men and women talking about being kidnapped, sexually abused, beaten, and worse. One guy had been strapped down while someone had carved him up. The counselors seemed to think this was where I fit in.

But ultimately, like the mental health clinic at college, the folks at Hazelden said, "We're not equipped to handle someone like you." They wanted to send me to a military trauma unit.

"What? You're telling me I have the symptoms of a guy who was in a POW camp?"

"Yes, you do."

"C'mon, you're just making that up, right?"

"No, Mr. Hammond. We think it would be best."

I could barely wrap my head around the idea that I really belonged in a trauma unit, but more than that, I figured if I went there, I wasn't going to get out for a long, long time.

I wanted to get back to New York and *SNL*.

After twenty-eight days, on a Thursday, I was released. On Friday, I flew home to New York. On Saturday, I did a cold open as Dick Cheney.

I think I got drunk a couple of times after I got back, but eventually I got sober and stayed that way. I went to meetings every day when I could. No cutting, no drinking, no flashbacks, no nothing. I was living in the alien territory of hope.

CHAPTER TEN

I'll Show You Multiple Personality Disorder, Pal

I f you've ever met me—and I'm assuming you haven't—you'd know that I'm as far from professorial as a person with a college education can be. But I think the art, if you will, of doing impressions offers what parents like to call a teachable moment. Bear with me.

For starters, an impressionist is not a practitioner of Impressionism, which involves a talent with a paintbrush that leads to the permanent collection at the Metropolitan Museum of Art rather than a late-night sketch comedy show on network television. An impressionist, to put it in a nutshell, is someone who imitates the voice, pattern of speech, and gestures of others, although the voice is the key thing. An impersonator specializes in imitating just one person, like the guy who showed up at your wife's bachelorette party dressed as Elvis.

Impressionists were all over British television starting in the 1960s, especially a guy named Mike Yarwood, whose impressions of famous Englishmen were so convincing that when he invented catchphrases, people assumed they came from the people he impersonated, even when they had never uttered the words themselves.

In the States, the dominant impressionist in the 1970s was Rich Little, also known as "The Man of a Thousand Voices." He first came to prominence when he was a guest on Judy Garland's variety show in 1964 and did an impression of James Mason, her costar in *A Star Is Born*. He went on *The Tonight Show* and did Carson for Carson, and he was a regular on Dean Martin's celebrity roasts. He was perhaps best known for his impression of Richard Nixon, which may have hurt him in the long haul, as he was still doing Nixon when he appeared at a White House Correspondents' Dinner in 2007. But Little was extremely versatile, and so good doing other people's voices that when the British actor David Niven was so frail that his voice wasn't strong enough for his roles in *Trail of the Pink Panther* and *Curse of the Pink Panther*, Little dubbed the voice and no one was the wiser, not even Niven, who didn't find out about it until he read it in the press. (I would do the same for longtime *SNL* announcer Don Pardo—one of the most recognizable voices on television—when he was down with laryngitis on the episode hosted by Elle MacPherson. And you know he had to be really ill, because Pardo still announces the show, well into his nineties and several years past his official retirement from NBC.)

Then Dana Carvey hit the airwaves in the mid-1980s, and no one was sure who the real George H. W. Bush was anymore. Even the former president found himself quoting Dana's impression of him in his eulogy for Gerald Ford in 2007 when he said, "Not gonna do it. Wouldn't be prudent." That's when you know you've arrived, when the person you imitate does an imitation of you imitating them. If I can get Bill Clinton to bite his lip and give the thumbs-up at the same time, I'll be golden. Dana would do something else I'd never be able to compete with—he did both Bill *and* Hillary Clinton.

Around the turn of the millennium, I had been invited to join the game show *Hollywood Squares* for a couple of weeks. In varying incarnations, *Hollywood Squares* was on the air for nearly four decades, starting back in the 1960s, when Peter Marshall hosted and Paul Lynde was the center square. If you're too young to have seen *Bewitched*, even in reruns, he was a delirious Uncle Arthur to Elizabeth Montgomery's

Samantha. And he won two Emmys for his center square gig over the course of a decade on the show. By the time I was invited on, the show had been revived by King World Productions, the fine people who bring you *Wheel of Fortune* and *Jeopardy!*—the real one, not the parody we did on *SNL*—and they'd enlisted Whoopi Goldberg as a producer as well as the center square.

The show had a long history of catering to the relative expertise of panelists. Paul Lynde was always asked questions that allowed him to make a snarky, and occasionally perverse, reply. Apparently *Planet of the Apes* star Roddy McDowall was a wiz at all things Shakespeare. Setting the precedent for me was Rich Little, a frequent *Squares* guest in the 1970s. Every time a contestant selected him, the question the host posed was usually about some celebrity or politician so that Little could do an impression of the person. When I was on the show, host Tom Bergeron did the same thing: every question—by amazing coincidence!—seemed to require my doing an impression in the answer. Over the course of a couple of weeks, that was a lot of impressions.

I was extremely flattered when Whoopi called me a shape-shifter. That might sound a little woo-woo mystical, but truth be told, it might actually be the best description of what I do. According to mythology, though, a real shape-shifter has trouble returning to his original form after a while, whereas I, unfortunately, had no such issue. Anyway, it was an amazing compliment.

Lorne recognized that predilection in me from the start, which is why I usually wasn't part of the sketches unless an impression was required. I think Lorne wanted me to focus on being ready to learn a new voice at a moment's notice. If one of the writers gave me an impression on Friday night, could I learn it by Saturday morning? Sometimes a sketch that featured Dick Cheney on Friday had been reconfigured to include Geraldo Rivera on Saturday. Lorne wanted me to be able to make that switch without the interference of having to prepare for other sketches during the week.

That's why over nearly a decade and a half on the show, I performed only a handful of made-up characters but more than a hundred impressions. From my first season to my last, the best known were Clinton, who I did more than eighty times, Dick Cheney (twenty-six times), Chris Matthews (twenty-two times), Al Gore (twenty times), Ted Koppel (seventeen times), Sean Connery (eighteen times), Jesse Jackson (fifteen times, including once as a cartoon), Dan Rather (thirteen times), Regis Philbin (twelve times), Arnold Schwarzenegger (eleven times), Donald Rumsfeld (ten times), and, in the year between March 2008 and March 2009, John McCain (eight times). But I would take on anybody they gave me, from Don Imus to Cincinnati Reds owner Marge Schott, one of two (notably unattractive) women I played. In fact, when I went out as Schott, they dressed me in a hideous short-sleeved shirt that clearly showed my scarred arms.

Doing an impression on *SNL* was a little like painting a picture while dodging bullets. Sometimes the paint gets smeared, sometimes the strokes are too broad, sometimes you don't even have time to put the brush to canvas.

I'd stand over whichever writer was conceiving my bit, doing the voice as he typed so he could hear it play. That goes on until all hours, and is why it is so unnerving to have someone say, "Okay, you're out of time. It's eleven o'clock, you gotta go." I always needed another hour, another week, another year. I never felt prepared.

If you look at the great broadcasters in history, these guys would not have gotten an A in the formalities taught in an oral interpretation class, but they're some of the greatest communicators of our time. From Walter Cronkite to Peter Jennings, Barbara Walters, and Howard Cosell, each had a unique way of speaking that resonated with the audience. "Idiosyncrasy" is the operative word, and it made each one memorable in his or her own way.

My challenge was learning to break each set of idiosyncrasies down: Where's the guy from? How old is he? Does he have a dialect? Is his voice high or low? Where does it originate in the throat, the back or the front? Phil Donahue and Ted Koppel are the same voice, but Phil's is a little bit more up front, and Koppel is in the back of the throat. Hopefully the subject has speech irregularities, something unusual about the way he forms words, the way his teeth and tongue work. Some people lose their accents over time, or they've corrected a speech problem, which doesn't really help because the world doesn't know that. I'd break it down into four or five factors, and I'd pick out a couple of hand gestures. I had to find a way to compress an entire person into a handful of tics and gestures.

Most of the people I did were in the news, so there was usually tape I could study. That wasn't the case when I did Clark Gable in a sketch called "Vincent Price's Thanksgiving Special 1958" alongside host Eva Longoria as Lucille Ball, Fred Armisen as Desi Arnaz, Horatio Sanz as Alfred Hitchcock, Kristen Wiig as Judy Garland, and Bill Hader playing Vincent Price. I'd never done Gable. Who the fuck was Clark Gable?

I'd seen a couple of his movies, but there was no reason for me to have learned Clark Gable in my whole life.

In the history of the show, Gable had been done only once before, by Dan Aykroyd in 1978. Somehow I had to find a way to Gable-ize the dialogue. I came up with a version that went pretty well in dress rehearsal, but the writers changed the sketch at the last second, and when I stepped onto the stage, my entrance was accompanied by a blast of loud violin music I wasn't expecting, and that threw me. I got the lines out, but I was disappointed with how it came out.

When I was lucky, I would get a new assignment that was similar to one I'd done in the past, which I could simply adapt. That cuts a lot of the time out of the process. I once tried to do Robin Williams, but I didn't want to portray him the way other people had in the past. The piece involved some dancing, which I hadn't done since the days of the B52s on those Disney ships. Stupidly, I spent hours trying to learn the dance moves instead of working on the voice. The impression was a total failure. Fortunately, the effort never saw air.

When I did Regis, I did him a little differently than Dana Carvey had. When I did Phil Donahue, I did him a little differently than Phil Hartman had. There was one sketch after Will Ferrell had left the show when I played George W. Bush, and there was a lot of pressure on me to mimic the way Will did him. But it wasn't possible; we're not wired the same. We went through that a lot over the years, where someone might say, "C'mon, just watch so-and-so do it." I never felt good about that, and I didn't do it.

Thank God I had the benefit of Karen Giordano's help in preparing to do Bill Clinton. She's a fantastic actress who also works as acting consultant. She's coached some of the biggest stars in the universe— she worked with Mariah Carey on her role as a social worker who uncovers the abuse inflicted on the pregnant teenage girl at the center of *Precious*—so I'm doubly lucky that she agreed to take me on. And Karen had me do the impression so that I was really doing Clinton. We gave Clinton a whole biography. I wanted to be funny, but I wanted to be as real as could be. Usually, the point was to do an exaggeration, but with Karen's help, I was going to do it as true to the man as I could. In actor-speak, most of my impressions *represented* the person, but my Clinton simply *presented* the person.

I worked hard to get Clinton right. I studied tape after tape of him. The average speaker has maybe five consistent hand gestures—Clinton had *thirty-seven*. My Clinton tapes were labeled "morning," "noon," and "night," because his voice changed toward the end of the day. Clinton loved being Clinton, he loved being president, every day was a huge event for him, and he got tired, which you could hear in his voice as the day went on and he grew raspier and raspier. Lots of people model themselves after other people: politicians do it, athletes do it, comedians do it, I do it. I noticed that Clinton seemed to be emulating John F. Kennedy's cadence, using commas in illogical places. The way I learned him, the way he finally came to me so that I could feel him, was doing Kennedy's inaugural address in a southern accent. Suddenly it all made sense to me.

Then when I got to meet him that first time in the Oval Office, there was his big red face, eyes a color blue that exists nowhere else in nature, and that little crackle in his voice. I call that the bedside vigil voice, the kind of gentleness that you would show to a dying family member. Clinton makes people feel good. He has a way of looking at you as if you were one of the most important persons who had ever stood before him. He did that for everybody in the room, in different rooms in different parts of the world for eight years. All these people walking away thinking, Wow, what the fuck was *that* all about? Who *is* that fucking guy? One reporter told me that in Washington they called it The Thing (He Does).

A few years ago, I was doing a show at a college. Afterward, when I was walking across campus, a coed approached me and said she'd flash me if I did Clinton for her.

"That is so sick," I said. In Clinton's drawl.

I thought, God, that was cheap. Did I just do that?

She was true to her word. Not bad. (I wish I were a better person, but apparently I'm not.)

One time I accidentally set my Hell's Kitchen apartment on fire. Note to self: lit cigarette in an ashtray on the bed plus oscillating fan and a quick trip to the deli leaving the aforementioned unattended equals bad idea. When I returned to the building, it was surrounded by fire trucks. I stood on the sidewalk and watched as my apartment was trashed first by flame, then by water.

One of the firemen noticed me lurking and came over.

"Sorry about your place, man. At least nobody was hurt."

"Thanks."

I was still looking at the wreck in front of me, wondering where I'd stay that night. Out of the corner of my eye, I noticed the fireman was kind of staring at me.

"Hey, are you the guy that does Clinton on TV?"

"Uh, yeah." *Shit, really?*

"Would you do a little Bill for us?"

Of course I did. The truth is, I was pretty flattered.

In 1999, as a new presidential race began, I started working on Al Gore, since he was Clinton's heir apparent for the Democratic nomination.

Gore was confounding because he spoke in three different ways. This wasn't like Clinton's morning-noon-night voices; these were three completely distinct ways of talking. How was I going to present *that* shit? I labeled my Gore tapes 1, 2, or 3, for each of the debates Gore had against senator and former NBA star Bill Bradley for the Democratic nomination. A fourth tape I used was of Gore at a backyard barbeque talking about how to cook meat in a safe way to avoid airborne viruses. I watched those tapes over and over.

When I was watching Gore, I knew he was not being himself. You couldn't argue that his style had evolved naturally, as the debates occurred pretty close together: the first in New Hampshire in January just before the primary there; the second in late February at the Apollo Theater in Harlem; and the third barely a week later, on March 1, in Los Angeles. The only theory I could come up with was that he had three different vocal coaches, which I guess you could say left him overcoached.

The rule in comedy is what's most personal is most general. You're more interesting if you're authentic or appear to be authentic. Gore said great words, but a lot more people would have gone for him if they'd seen something different, if they'd seen the real man. I saw a photo of him, sweat on his brow, tie loosened, a Heineken in his hand after dancing with his wife, and I thought, If *that* Gore had campaigned, wow, what would that have been like? In person, he was engaging and funny, but that wasn't what the public was seeing.

When Gore won the Nobel Prize in 2007 for his work on climate change, he struck me as a real phoenix rising from the ashes. Notably, he didn't speak in that awkward, overcoached way anymore. He was the guy whom I had seen privately. He was utterly himself. When we did a sketch about his winning, I wanted to play him like that, but theatrically there was a concern that the audience would be expecting to see the old Gore. I asked Lorne what he thought. He said, "If you think you can split the difference between the new Gore and the old three Gores, do it."

I went out there and tried to do the four voices at once. It was 50 percent of the real guy, and 50 percent of the guy from before. We were prepared to sacrifice accuracy if we thought we could get a laugh, and we did—the sketch garnered three applause breaks—but it wasn't pleasing to me or to anyone else to play this new fellow, and we never did it again.

Given the nature of my craft, I study people even when I have no expectation of ever doing them. I learned a lot about the world leaders just by watching them. For instance, it was clear to me that George W. Bush couldn't act and didn't want to. I think there was nervous anxiety about being in the presidential spotlight, and sometimes he'd say things the media would make fun of. He gets tripped up on shit that doesn't interest him. He's not good talking about wetlands because wetlands are not his thing. However, he *was* good when he talked about the war on terror. The point is, Bush in person is exactly what you see on TV; he's the same guy. Clinton too, same guy. McCain, same guy. Sarah Palin, same woman.

And then there was Ronald Reagan. He understood language better than any creature that I've ever seen. There was a brilliant sketch with Phil Hartman doing Reagan at a press conference. When the media left the room, Reagan became this brilliant, articulate guy. That sketch resonated with the entire world, because on some level people understood exactly what Reagan was doing. His understanding of the spoken word was astonishing. He could manipulate a press conference or a hostile press corps better than anybody. Reagan understood that the evening news could only put on a ten-second sound bite.

Also, Reagan lived and breathed the things he said. When Reagan said, "Mr. Gorbachev, tear down that wall," it came from his balls. You could hear his balls go *clank*.

I did him in my stand-up act, but when I was on the show Reagan wasn't much in the news, aside from dying from Alzheimer's, so there was never a call to do him on air.

There's a benchmark that you go for at *SNL*: Is the impression good enough to be funny? If you're getting laughs in dress rehearsal, it's going to be on air. Lorne put me up as Gore maybe two or three times in 1999, but the sketch didn't get picked because it wasn't funny. Nobody believed me when I said, "Trust me. This is how the guy talks." So I studied and studied and studied. I still have photos of the very peculiar ways in which his mouth would form consonants and vowels. I finally did him on air in late '99 and early '00, but it still wasn't clicking. He just seemed like a big guy with a southern accent. I tried out different versions at the Cellar for about a year before I found one that resonated with the audience.

I added in that when he turned left or right, it was as if his upper body were on an axis because the neck wouldn't turn by itself. The shoulders had to turn when the head turned. I also noticed that he occasionally employed, as countless politicians have, a slight Reaganesque head wag, so I used that too. I picked a couple of different laughs. Everyone laughs without really realizing it, and they do it in four or five different ways, but for an impression it's best to stick with one or two.

It wasn't until the debate sketch in the cold open of the season premiere in October 2000, a month before the election, that my Gore, an amalgam of all three of his voices, really hit (even though I was far too out of it at the time to notice right away).

Usually, I didn't have the luxury of that much intense study time. Even if a writer brings you a voice on Wednesday, you don't have forty-eight hours to master it, because you're prepping other things simultaneously. I'd be lucky to get in five or six hours to work on it. It's also really hard to absorb more than an hour or two a day of a person. I would frequently study someone intensely for a couple of hours, and then try not to think about them for the rest of the day. I'd go back to it the next day and see if some particle of them had imprinted on me.

I was starting to perfect that technique in my first year on the show in a sketch with Norm Macdonald's wildly popular Bob Dole where I played Speaker of the House Newt Gingrich. From a technical perspective, Gingrich's voice was high-pitched but very resonant, and he had a very interesting dialect, but Gingrich was known more for what he said than how he said it, so I raised the voice just a little bit, tucked it a little bit in the back of the throat. I didn't want to go too far back, or it would have been Kermit the Frog. Then I tried to find a vowel substitution. I think his "eh" sound became an "uh" sound. I posted that vowel substitution in my bathroom and in my kitchen. Unfortunately, the resulting impression wasn't anything to write home about, and I never did it again.

Half the time I left 30 Rock on Saturday night thinking I had disgraced myself. I hated that we were always on the clock. As Lorne says, the show goes on because it's eleven thirty, not because it's ready. Everyone is held to such an incredibly high standard by the media, but they don't realize we're putting those things together in hours. *Hours.* I would be so tired and so confused from having tried to master God knows how many voices in one week. The clock is ticking, and you're constantly trying to work with new material and then look back at your tapes to see if there's something you're missing.

A couple of times while I was there the writers presented me with an altogether different challenge: talking to myself playing another character. In one sketch, I was live as Ted Koppel interviewing myself on tape as Clinton. The only problem with that is, I couldn't tell how long to hold for laughs after a joke line. I'd be in the middle of a paragraph, say a funny line, and the laugh would cover up the next two lines.

And once, the voice I was doing got screwed up in a completely new way. When Queen Latifah was hosting, we did a Regis Philbin sketch. I always played him a little over the top, but this was different. I'd taken steroids earlier in the day because I was having trouble sleeping and I was exhausted. Two little pills, and for sixteen hours I was so finely tuned and so springy, I was like a shiny new toy. My whole being felt stronger. But my brain and mouth had trouble coordinating all this new energy, and my speech seemed to take on a life of its own. When I got to his trademark "Am I right, Gelman?" part of the sketch, I said it so manically that it was as though Regis had turned into Porky Pig.

I felt like the work, with the exception of Clinton, was always incomplete. It always amounted to a last-minute distillation of age, dialect, speech impediment, and, if I was lucky, a couple of gestures.

With Bush's secretary of defense, Donald Rumsfeld, I started with Tom Joad. That is, Henry Fonda as he played Tom Joad in the 1940 film of Steinbeck's *Grapes of Wrath*. It struck me that the two men had similar accents and tones. So I started with Fonda, and then made him older. The physical part was pretty easy to master because Rumsfeld held himself in such a particularly taut way. And his attitude with the press was distinctive; the way he held his head up and squinted, as though literally looking down his nose at the mere mortals assembled before him. Those were easy gestures to imitate. We introduced him on *SNL* in November 2001, in the wake of 9/11 when he was frequently in the public eye. In a sketch mimicking Rumsfeld's notoriously testy press conferences, I said, "I'd like to give you a better answer to that question, but I fell asleep during the first part of it."

So there you have it: the man who so frightened the press that they started to call him Dr. Strangelove was based on a mild-mannered Dust Bowl Okie, and a fictional one at that. I heard later that the secretary of defense liked my impression of him, although when asked about it by a journalist during a press conference, he snapped, "If I want to talk about *Saturday Night Live*, I'll bring it up." It was a great line, and, as I recall, it got a nice laugh.

Near the end of my first season on the show, the *Daily News* called me "the best impressionist any-where, at any time." The following week, the same paper gave me a shitty review about a bad impression I did of Tom Brokaw. I ran into the reviewer at a coffee shop a month later and told him he was right, and he was. I never did Brokaw again, happily handing those reins to Chris Parnell, who did extremely well with it for ten years, right up to the second presidential debate sketch we did in the autumn of 2008 when I played John McCain against Fred Armisen's Barack Obama, and Parnell, who'd left the show two years earlier, came back to play Brokaw as the moderator.

There isn't that much cracking up on air, but it's a guaranteed laugh from the studio audience if you sud-denly slip and giggle. I was pretty good at keeping it together, although it was a struggle whenever Amy Poehler's Kelly Ripa climbed all over my Regis Phil-bin. But I out-and-out lost it on a few occasions when something unexpected happened.

The first time was in my third season, when we were doing a sketch called "Riding My Donkey Political Talk

Show." It was me, Will Ferrell, Tim Meadows, Ana Gasteyer, and Jim Breuer. It was exactly what the title said it was, a political talk show that took place while riding donkeys. I was playing Sam Donaldson.

At one point in dress rehearsal, my donkey turned around and looked at me, then looked at my crotch as if thinking, "Excuse me, but are those balls by any chance? I'd really like to find out." And went for it. The sketch fell apart, but—assuming we could get the donkeys under control—it showed a lot of potential.

By the time we got on air, the donkeys had been sedated or something. I was wearing a catcher's cup, just in case. When my donkey whirled his head to see if my crotch still existed and tried to make a go for it, his legs went out from under him. We all started laughing. All the donkeys were having trouble keeping their footing. I think it's the funniest thing I've ever seen anywhere.

The next time, in March 2002, was during one of my favorite sketches ever, with Horatio Sanz and Sir Ian McKellen. Horatio was genius playing a Turkish talk show host, Ferey Muhtar, and in a rare instance where I didn't do an impression, I was his Turkish Ed McMahon, Tarik. Ian McKellen played a Turkish dance club owner who was a guest on the show. All three of us were wearing dark wigs and mustaches, and everybody was smoking cigarettes.

It was funny enough when Horatio started the sketch blathering in faux Turkish slang, and then Sir Ian came on and kept touching his package, which cracked the house up. I didn't realize that my mustache was coming off, and every time I spoke, it flopped around my face. As I'm speaking the line to Sir Ian, "Don't give me that suck job, man," he reached over and tried to reattach my mustache. I tried to stay in character, but the audience was howling, and I lost it. Horatio giggled every time he looked at me—the reattachment failed, so the mustache kept flapping. Finally Horatio ad-libbed, "We all know you got a drinking problem, man, c'mon! You got the fake 'stache! You can't even grow a 'stache, man!" By the end of the sketch, the mustache was gone, Horatio—staying in character by the merest of threads—was saying "We've got a show to do, bro!" and I could barely speak my lines, I was laughing so hard.

The third time happened during a cold open in 2004, a re-creation of the last episode of *Friends*, but with Secretary of Defense Donald Rumsfeld leaving the Bush White House instead of Rachel leaving Ross. I played Rumsfeld opposite Will Forte's Bush (he'd taken over the recurring role after Will Ferrell left the show). I remember the script was still being written as we were approaching the stage, so I didn't have a handle on all of it. A lot of it was shown to me on paper, but I hadn't seen the cue cards yet. It didn't really matter, because the house exploded into the loudest laugh I ever heard when Rumsfeld and Bush embraced in a big full-mouth make-out. As we unclasped, I turned to the camera to say "Live from New York," and Will snuck in two extra kisses to the side of my face that weren't part of the script. That was the only time I was fully facing the camera when I cracked.

When I walked offstage, Jeff Zucker said, "Classic." He seemed very pleased.

The last notable time happened in 2007, when Molly Shannon came back to host the show. Molly is a genius at four-minute theater, and she always held something back during rehearsal that she'd spring on us on air. No one has ever consistently blown the roof off a place quite like she did.

During this show, we did a sketch where I was Dan Rather moderating a presidential debate between fringe candidates that included Bill Hader as Tony Blair, Amy Poehler as Dennis Kucinich, Maya Rudolph as Fantasmagoria Purlene Robinson, Andy Samberg as Lord Simon Frothingham, and a bunch of others—you can imagine how bizarre it was.

When it was over, I had three minutes to change from staid newscaster Rather to Jersey gangster Tony

Soprano for a sketch in which Tony auditions a new dancer for the Bada Bing club. I was backstage in one of the change booths under the bleachers—where we went when there wasn't even time to go to the quick-change booths down the hall—while someone from hair pulled Rather's wig off me and then squeezed an almost-baldy wig on; someone from wardrobe helped me out of the suit and into a bowling shirt and slacks; someone from makeup took one nose off and put another on; and someone else from makeup dabbed me with foundation. If I'd had time, I would have gone over to the cue card department, under the bleachers on the other side of the studio, to have at least a quick look at the final dialogue, but there was no time. In the middle of this, a piece of scaffolding fell on my head.

By the time the three minutes were up and we went on air, my eyelashes were semi-glued together, I had pancake in my mouth, and I was trying to get the sting out of my eyes. I looked up at the cue cards and I didn't recognize the dialogue—at all. I was so thrown by everything that had just happened that my Tony voice came out like Brando with a New York accent, which, as it turned out, worked well enough to get the laughs, but it was not what I had planned.

Then Molly came out. She was playing her recurring character Sally O'Malley, who always proudly declaimed, "I'm fifty years old!" With an unflattering red knit pantsuit and a white handbag slung over her arm, Molly attempted to dance provocatively around the stripper pole. I interrupted her "moves" to tell her she was too old, but nobody puts Sally O'Malley down: Molly hiked up her pants halfway to her chest to reveal the *biggest camel toe ever seen on live television.* Molly pointed to her "desert rose," then put her foot on the top of my head, forcing me to look right at her crotch. The place was complete thunder, and I was gone.

CHAPTER ELEVEN

I Saw What You Did, and
I Know Who You Are

I f I've learned anything on this earth, it is that people will pay money to laugh. Recession or no recession, war in the Middle East or tornadoes in the Midwest, the Comedy Cellar is packed every Friday night. Say you're bored on a Friday afternoon, you want something to do, so you go see a comic do a sixty-minute show. On a good day, that's five laughs a minute—a laugh every twelve seconds for an hour. That's three hundred laughs. I think that's worth a two-drink minimum.

During the *SNL* season, from September to May, I went to the Cellar to work on new impressions Sundays, Mondays, and Tuesdays. I never went on Wednesdays, because that was read-through, and obviously I didn't go on Friday or Saturday (except for that late show at

Caroline's after *SNL*). During the summer hiatus, I hit the national circuit.

Once I was on the *LIVE! with Regis & Kelly* show, and I was getting ready to say good-bye when someone handed Regis a card to read. "Darrell Hammond will be at the Improv in Baltimore this week." He looked over at me with a cocked eyebrow. "Improv in *Baltimore?*" That eyebrow said, What in the world are you doing *that* for?

Truth is, I didn't turn many jobs down. I knew that one day *SNL* would end, and the chances of having a second career in show business were slim to none, so I wanted to make every single cent that I could possibly make. I had a lot of energy. I drank a lot of Red Bull. I'm not blowing smoke up their ass because I want an endorsement deal (although I do). I discovered that magic elixir on weekends when I would fly to Arizona or some other place on the far side of the country, and I'd get wicked jet lag. I'd do an hour of stand-up already tired as fuck, and still have to face a second show. The audience would be rowdy and noisy and not paying attention, so I'd have to be at my über-best to score and control them. A couple of Red Bulls, and I'd be crystal clear. I'll be honest, the best performances I've ever given were while riding that Red Bull wave. And I'm not saying that because I want a contract (which I do).

I don't know if it's comedy in particular, or fame in general, or maybe it's just me, but I seem to attract strange women. Not my wife, and not some of my significant others, but out there in the world, my molecules seem to cry out to the oddballs, "Hey, wanna hang out?"

I was sitting in the Cellar one night after a show when a girl came and sat down. She was in her mid-twenties, and had been watching *SNL* since she was a kid. She talked about the minimalist approach I took to Cheney, and the lyrical swoon of Clinton. I thought she was an English major. I was very impressed.

"What do you do?" I asked.

"I'm a fetishist," she said.

"What is that?"

"I jerk people off with my feet."

I should have won an Academy Award for keeping my composure.

"Can you make a good living?" I asked, as though she'd told me she did something as ordinary as trading pork futures on the Commodities Exchange.

"Oh, yeah. I go to their apartment or sometimes a hotel. We talk about Chaucer or Rembrandt for a while, drink some wine, then I jerk 'em off with my feet. I charge a lot of money for that."

Then she said, "I kind of have a crush on you. Would you be interested?"

A little alarm bell went off. No matter how cheery and well put together she was, something told me to sit this one out.

"You know what? Maybe another time." I've regretted it since.

Another night I went to a club on the Lower East Side with a runway model, and everything in the place was yellow. They had yellow plastic forks, yellow knives, and yellow plastic spoons, yellow napkins, yellow counters, yellow walls, and yellow cups.

There was a guy speaking French to somebody playing chess. His fingernails were really long and dirty, his hair was matted and speckled with dandruff. He started talking to me about a sketch that I'd done on *SNL*.

Meanwhile, this runway model, who was ostensibly my date, wandered off into the back room.

"What's back there?" I asked.

"Some of the guys back there will trade coke for sex."

Oh. Of course.

Sometimes the women were more than strange. Sometimes they were dangerous.

My second stalker was an amateur. A Turkish woman who showed up downstairs at the security desk asking to see me. She claimed my nonexistent brother was the father of her baby and she wanted money. She should have done a little more research on my family tree before she got to work, don't you think?

The first stalker, however, was quite accomplished.

I met this attractive, charming, well-spoken woman at the Cellar one night in the early years of the millennium. She was a fan, we hit it off, and I invited her back to my place.

When we got to my apartment, she said, "I have to use the facilities," and as she walked toward the bathroom, she turned and looked at me—it was as if someone else's eyes had been transplanted into her head. She might as well have said to me, "I saw what you did, and I know who you are." *Air raid sirens.*

"I'm not feeling well all of a sudden," I said. "I need to go see some friends of mine." I'd seen those eyes before, hadn't I?

I escorted her downstairs to a cab.

Later that night, she called me. I didn't pick up. She called again. And again. In total, she called about fifty times in the space of an hour. In most of the messages, she was screaming.

I took the messages to the local precinct. There was one detective nearing the end of a brilliant career who took care of it, but it took time, because, as he said, "This animal needs to come out of the bushes a little bit more. She's not going to be happy with just yelling at you. She wants more, and she's going to eventually have to come out to get it. So we wait."

He eventually called her and said, "You need to understand that this is not going to work out for you. It's going to be bad if you take this any further. I'm obliged to tell you that. I think it would be really great if you would just stop."

But he understood this kind of person. He knew she wasn't going to stop, and she didn't.

Her next move was to have a friend of hers call this veteran detective and say he was a lawyer, telling the police what they couldn't do. The boys at the precinct house just laughed.

Then she sent a horrible note threatening my family, and that was enough. The detective and I went to the district attorney, and the DA said, "If we want to take her to trial, will you testify?"

Ninety-nine point-nine-nine-nine percent of the time, once the DA and the NYPD get involved at the level they were prepared to get involved at, the harassment stops. Her friend got a call from the DA's office, which was sufficient to change his mind about the things he'd said to the cops. Then the police went to her apartment, which I was told was done up like a shrine to me, with photos of me and interviews I'd done plastered all over the walls. They arrested her and incarcerated her at either the Tombs downtown or Riker's Island. As the cops suspected, the experience was sufficiently horrific, and it was no longer worth it to her to continue. She copped a plea, and I never heard from her again.

And yet, although they stopped her, they couldn't stop the fear I'd developed. It's hard to do comedy when someone's threatening to kill you and your family. When someone has a gun and can shoot from the shadows, they might as well be a hundred feet tall—either way, there's nothing you can do to stop them. They've let you know that they want to do you harm, they let you know they know where you are, and due to the stalker laws, they have to make some kind of threat on tape or on paper before the authorities can get involved. But even then, how much can they do if a person really wants to kill you? Put them in jail for a few days?

It reached the point that I was leaving 30 Rock via the garage in the subbasement so that no one ever saw me exit the building.

I didn't want to stop going to the Cellar because it was really helpful to me to work out new material there. Eventually I called Estee and said, "I need to do a set, but you can't advertise that I'm gonna be there." Sometimes my friend Eddie Galanek—a former NYPD detective who'd spent a couple of years undercover with the Gambinos before he started doing security for *SNL*, scanning the audience for troublemakers and looking after the celebrity hosts—would come with me, but you can't get a top bodyguard to come with you every fucking night.

Still, you've got to be security conscious. You never know when you're going to encounter some freak who was hit on the head when he was three years old at vacation Bible school after his dad molested him and spent the next ten years chained to the radiator in his grandma's basement. It's true that 99.9 percent of stalkers are adoring fans, but 0.1 percent is the guy chained to the radiator in his grandma's basement, and that guy wants to kill you.

CHAPTER TWELVE

A Host of Hosts

For the hosts, being on *SNL* is a mixture of pampering and really hard work. They get lots of encouragement, lots of support, and there's always an assistant or two floating around to make sure they have everything they need. There were some extremely famous people who walked onto that stage before, during, and after my tenure, but live television is a very different—and more daunting—experience from what most Hollywood celebrities are accustomed to. There are no retakes. You screw up, and millions of people will know about it. (Just ask musical guest Ashlee Simpson, who got caught lip-synching to the wrong song and walked offstage humiliated in 2009. If that weren't bad enough, *60 Minutes* was on set that week, so the fiasco was replayed for a second time to a brand-new audience.)

For the most part, people are on their best behavior as they walk through the hallowed halls of *Saturday Night Live*. No one wants to feel like they're going out in front of sixteen million people alone.

Strangely enough, it wasn't the showbiz people who seemed to have the easiest time of it. The athletes just expect that they're going to figure out a way to do this. All of these hall-of-fame superstars would walk in seeming to think, Yeah, this is how I do things—I win. Tom Brady was fabulous. Peyton Manning was amazing. Jeter was great. According to the security guys—and they knew dirt on everybody—Jeter is one of the classiest guys going. He certainly behaved that way around *SNL*. Sometimes while raising my kid, I'd think, What would Jeter do? (When I did Trump, as I left the stage, I would say, "And Derek Jeter's going to be there," as a tip of the hat. Jeter's Mr. New York City, as far as I'm concerned.)

Donald Trump was the same way. Trump doesn't quite understand defeat. Even when it occurs, it doesn't register in his brain. He has been millions in debt and never developed a drug problem, like some people would have. He reminds me of the horse Native Dancer, one of the most famous thoroughbreds in racing history. Back in the 1950s, he was undefeated for three years running, and he became accustomed to going into the winner's circle after every race. When he finally lost the Kentucky Derby, he trotted over to the winner's circle anyway. Winning was all he knew.

It reminds me of an old soft drink commercial where Larry Bird was supposed to miss a free throw. *I'm not being guarded, the basket is only ten feet away, and you're asking me to miss this shot? I don't know how.*

When Trump hosted *SNL*, he dove in with tremendous enthusiasm. Here was one of the most famous men in the country asking more questions, spending more time with the cameraman, the lighting people, the costume people, than just about any other host I'd ever encountered.

At one point, shortly before we went on the air, he said, "Let me ask you a question. If I deviate from the script for a second, will that make things seem more spontaneous?"

Of course. Nothing could be more spontaneous.

"And if it's more spontaneous, it's more real?"

Yeah, and that makes it funny.

He'd been studying this for less than a week, but he broke that down pretty fucking fast. I was amazed, especially considering the chaos with hair and makeup and costuming that was going on backstage in the seconds before I was to join him on the main stage during his monologue:

"Where's the brush?"

"Nobody said anything to me about a brush."

"Did you get me a nose?"

"Didn't have time. Just found out."

"So did I. Did you get the elevated shoes?"

"Why do you need elevated shoes?"

"Because he's a fucking foot taller than me."

"Okay, tell Tom to hurry."

"How's the wig?"

"I didn't get a chance to fit it."

"Why?"

"Nobody told me."

"Fuck."

"Here are the shoes."

"Thanks. Is that the wig?"

"Yes."

"That doesn't look like an onion loaf."

"What?"

"His hair is supposed to look like a fucking onion loaf! It's fucking Donald Trump. Jesus."

"Oh, no."

"What?"

"Oh, fucking no!"

"What?"

"These shoes don't fit."

"Ten seconds. You'll have to go on without them."

"But he's a fucking foot taller than me!"

"Jesus, calm down."

"I can't. I'm doing Donald Trump to Donald Trump, and he's a fucking foot taller than me."

"Deep breath. Two seconds."

"I've got no fucking shoes on! No onion loaf!"

"You're on!"

The bit worked terrifically well—Trump had me "fire" Jimmy Fallon's wildly gesturing Jeff Zucker—but in our matching dark suits and purple ties, I ended up looking like a Mini-Me version of the dude instead of his double.

Shaquille O'Neal was a little different because he had someone with him who appeared to act as his adviser. I happened to walk by the music room and saw this giant man sitting on a tiny couch with a normal-sized man sitting next to him, while writers and music people pitched a song they wanted him to sing. Without a word, he looked at the guy next to him, who must have said or gestured his approval, and then he said, "Yeah, okay."

Sometimes, when you're that famous, you've got to have someone who can say, "You're not going to look good if you do this."

Of course, Shaquille O'Neal singing a lullaby while holding Will Ferrell like a baby was a smash hit. Also, he ate more in one sitting than a mere mortal could eat in a year. Nice guy, real soft-spoken.

When football stud Tom Brady hosted in 2005, Robert Smigel wrote a TV Funhouse sketch called, "Sexual Harassment and You." The bit required Brady to walk through a warren of office cubicles to show how to "correctly" talk to women. In one scene, he walks through in a pair of tighty-whities and cups Amy Poehler's breast. At read-through, someone had marked at the top of the script, "This one is for the costume shop."

We also did a Dr. Phil sketch in which Tom played a husband who's so clueless to his wife's needs that Dr. Phil hauls off and slaps him. I was very proud of that because he demonstrated his athleticism by lurching his head with impeccable timing as I whizzed my hand past his face.

J. Lo was one of those hosts who doubled as the musical act. When she appeared in February 2001, I got to see her extraordinary magnetism up close. Just walking down the hall, she owned the place. I overheard some tech guys by the coffee urn say, "Animals probably look at her and think, Why don't *I* get that?" She's certainly an attractive woman, but this is *Saturday*

Night Live. If you're hot, we've seen you. J. Lo was different. Catherine Zeta Jones had an imperial kind of beauty—you'd want your queen to look like her—but J. Lo has the shit that people go to war for, and this was a full ten years before the country fell in love with her all over again when she cried for the contestants as a judge on *American Idol.*

There were men who seemed to have that effect on women too. I remember walking down Fiftieth Street with Scott Wolf, the young heartthrob from the hit drama *Party of Five,* and a convertible filled with attractive young women pulled up alongside us.

"Hey, Scott! Come with us!"

Is the balance of nature upset when something like that occurs? There's no wooing, no cooing, no courting, no flirting, no getting to know you, no sizing each other up, no smelling each other's cologne, just "Hey, Scott! Come with us!" What is *that* like?

Shortly before air on October 18, 2008, I stepped out of my dressing room and slammed into a wall of the tallest, largest badass Secret Service agents the U.S. government has ever assembled. Before I could ask what was going on, I saw Sarah Palin walking past the hair and makeup department, past Jason Sudeikis's dressing room, straight toward me. Wait, no, not toward me—toward the bathroom. Only the host and musical guest get a dressing room with its own bathroom. Other guests who make cameos, as Palin was to do that night, are consigned to public facilities.

Oh, no.

I drank tons of Diet Coke when I was working, so I used that bathroom a lot. The truth is, I'm pretty good about the bathroom, ladies. I lift the seat, and my aim is true. Most of the time.

I can't let her go in there. What if I missed? I have to check! How do you say that to the Secret Service?

I could just picture it:

"Is there a problem here, son?"

"No, sir, but I need to get in there for a sec before she—"

"Stop right there!" Guns drawn.

"But, sir!"

"Don't move!"

"But I sprinkle when I tinkle!"

I once saw James Gandolfini go in that bathroom and I thought, Eh, he's okay. In fact, I was secretly charmed that Gandolfini would use the same facilities that I had been using that day.

But with Sarah Palin, I was a little alarmed. I felt she could have made a better choice at Rockefeller Center. I would have advised her to use the bathrooms down the center hall that leads to the stage doors and the theater. Those are very fine bathrooms, very well kept. I think that's where most celebrity guest star people go to the bathroom. I wanted to say "Hillary doesn't use that bathroom. Please let me try to discourage you. Hillary uses the other one."

An hour later, someone asked me, "Would you like to say hello to Mrs. Palin?"

I walked down the hall to her dressing room. It was a peculiar moment. You know when your mouth gets ahead of your brain and a thought rolls off your tongue as your brain says, "Hang on a minute! I did not authorize that! Stop!" Well, when I was with her, I didn't know what to say. "Hey, how'd you like the bathroom? Everything okay? Find everything in order?" Somehow I managed to keep that witty repartee inside, but I can't remember what I actually said, because all I could think about was her using the toilet I peed in all day.

I'm not speaking for or against her politics, but she seemed a little bit like royalty in the degree of attention she commanded. The members of her security detail I spoke to said they'd never seen this kind of preparation for *anybody*. The NYPD was out in force. God knows what kind of sniper SWAT team was stationed on top of the building. I've never seen so many Secret Servicemen in my life. More than for John McCain, more than for Barack Obama, more than for Hillary Clinton, more than for Bob Dole, more than for anyone. When Sarah Palin came, we weren't allowed to go into the studio to start work that Saturday morning until they'd searched all of our dressing rooms.

At the same time, Sarah Palin is exactly what she seems. She's not putting on an act. That's who she is. She's incredibly charming, and she's very attractive. If you meet her in person, and you claim not to like her after five minutes, I don't believe you.

When Illinois senator and presidential candidate Barack Obama did a cameo in November 2007, someone had just discovered he was a distant relation to Dick Cheney, so the writers came up with a terrific bit about it that he and I would do for "Update" in which I would play Cheney, and we'd chat about our family. On Saturday afternoon, we took it out on the floor for the run-through. On the floor, you can always improvise a line to see how far you can stretch. If the writer likes your ad-lib, he might include it in the final sketch.

When we did the Cheney/Obama bit, everybody in the studio laughed, but Obama decided he didn't want to do it. He apologized, but it was killed before we got to dress. Instead, we did a cold open with Bill and Hillary Clinton hosting a Halloween costume party. I was Clinton, dressed as a character from the show *The Pickup Artist*, and Amy Poehler as Hillary was dressed as a bride. Horatio Sanz, as Governor Bill Richardson, sucks up to Hillary in a not-very-subtle attempt to win a spot on her ticket as vice president.

When he wanders away, Hillary says to Bill, "How come people are only nice to me when they want to be vice president, and they're nice to you all the time?"

I reply, "People like me."

A few months later, when Hillary was on the show, I wondered if she thought about that line. But then again, I've been to twenty-three shrinks, what do I fucking know?

In walks a guy wearing a Barack Obama mask. Amy/Hillary asks, "Who is that under there?"

He takes the mask off, and lo and behold, it's the actual future president (although of course we didn't know the future president part yet). The guy had so much charm and charisma, he might as well have been Jack Kennedy, and he was greeted with such loud applause and cheering from the studio audience that we had to pause before he could yell, "Live from New York!"

When people like Obama and Palin show up for a cameo like this, rather than hosting, they aren't there the whole week. They're too busy kissing hands and shaking babies—all that important stuff that constitutes running for national office. So they show up on Saturday. It gives you an idea what a quick study a presidential candidate in this country has to be. You've got to be able to think on your feet, absorb a ton of material really quickly. You can see why a guy like Bush, who's not really interested in being a performer, can stumble around a bit out there, and get in trouble once in a while. But when Bush came and did the presidential special, he did some nice work.

Bob Dole had impeccable comic timing. He did a cameo right after he lost his bid to unseat Bill Clinton in 1996. He was in the cold open with Norm Macdonald, trying to console Norm since, because he'd lost, Norm's impression of Dole wasn't necessary anymore. Dole nailed his punch line: "You're really doing an impression of Dan Aykroyd when he does an impression of me. You know it, I know it, and the American people know it."

I remember thinking, How's he so good at this?

When Arizona senator and war hero John McCain hosted in 2002, he easily mastered his monologue. We did a sketch of "Meet the Press" on that episode where he played himself, and I played the late great Tim Russert. I have to say, McCain knows how to wait for a laugh; he's got great comic timing. He came back for a cameo on the last show before the 2008 election, in which he and his wife, Cindy, were in the cold open with Tina Fey as Sarah Palin. The three of them hosted a QVC "Washington Outsider Jewelry Extravaganza" to raise money for awareness of the campaign. They were fantastic. It had been years since I'd been onstage for the good-byes—in retrospect, it was an asshole thing to do, but I never really felt like a full cast member, so I usually just left when my last sketch was over—but I did it when McCain was on the show, out of respect for him.

When Hillary came on in the spring of 2008, there was a palpable buzz in the air about her. I've been around the most famous and powerful people on the planet, and some of them have a certain *thing*. When Hillary walked into *SNL*, she was a star. She walked down the center hall toward the studio twin doors with the magnetism of a president. Everyone was fascinated by her, whether they loved her or hated her. Young writers, young cast members, people who were planning on skewering her that night, were all like "Oh my God." She looked like a head of state.

She did a bit with Amy that required her to absorb material in a really short amount of time, and it went extremely well.

She was also very friendly and forthcoming with the cast and crew, and lots of people were charmed by her. As for me, well, I think she may have been unhappy about that Halloween sketch I did with Obama. Or maybe it was my appearance at Bush's first Correspondents' Dinner in 2001, when I did a bit as Bill Clinton explaining to the men in the audience why he'd messed around with Monica Lewinsky: "You've met my wife?" Anyway, when the junior senator from New York passed me in the hallway that Saturday, she looked at me like I was a piece of furniture. I was upset that she seemed unhappy with me. You like to think the high and mighty are delighted by you too, but maybe the high and mighty don't give a shit about you, you fake-nose-wearing fuck.

The following Monday, I suspected I would run into her again, because she was scheduled to appear at an event I had to do for Lorne at the Plaza, during which I was supposed to give a pair of shoes to Mayor Bloomberg. Hillary didn't come in until I was leaving. I pictured her waiting backstage, saying, "Is that fucking asshole out of there yet?"

Back when she still liked me—she sent me an autographed photo thanking me for playing her husband after the first Correspondents' Dinner I did—Hillary invited me to come perform at an event dressed up like Bill. Amy Poehler was going to join me and play Hillary. Amy and I wrote what I thought was a gorgeous piece based on the musical number "If I Loved You" from *Carousel*: "If I Did Run." Amy sang the first verse, and I sang the second, and then we joined in together. Someone from Hillary's campaign pooh-poohed it. Unfortunately, it was a Friday night at *SNL*, not exactly a good time to stretch out with your feelings and think about other things. We didn't have time to come up with anything new. I'm pretty sure it was that Halloween sketch that ticked her off, but Amy and I were both honored that we'd been invited.

Having spent a fair amount of time studying these people, I had formed an idea of how the various politicians stacked up against one another.

The thing about Bush is, if you dropped him in New York City, and he didn't have a famous dad and he didn't have any money and he had no education, he would have an apartment in a month, although he'd have some bumps and bruises and scrapes.

Bill Clinton, in the same circumstance, would also have an apartment, but he would have a new wardrobe to go with it. Both were very canny, streetwise guys who knew how to navigate.

Obama would have the apartment, but he would also be teaching an advanced seminar in Eastern Lit at the graduate Department of Linguistics at Columbia.

Thanks to all the years my daughter has hung out backstage—she basically grew up at *SNL*—she pretty much thinks everyone famous will show up in her world sooner or later, because that's what her life was like. Twenty new superstars a year over twelve years of her life—that's a lot of celebrities. Obama had a private heart-to-heart with her and posed with her for a photo. I thought that was completely charming, and she wasn't at all intimidated.

When I saw Kate Winslet backstage at the Letterman show, my daughter was very small. Kate walked right up to her and said, "What a beautiful baby." She leaned down and gave her a little kiss, and then she picked her up, and my daughter tried to kiss her back. Kate said, "No, my darling, you must *purse* your lips. *Purse* your lips."

A few years later, Kate hosted *SNL*, and I reminded her of the incident. My daughter was about six or seven, and Kate said, "Remember what I taught you?" My daughter said yes, and they reprised the exercise. It was easily the most elegant, sweet moment I ever saw.

Even when my daughter met the Olsen twins, she wasn't that blown away. If you were on television, sooner or later you were going to show up in the room with her. That's the way it was.

I used to get hideously jealous of anyone who got something I wanted and didn't work for it. I practiced for eighteen years to get to *Saturday Night Live*. All Paris Hilton had to do was make a home sex tape that was so grainy and dark, like it had been filmed with night vision, that you couldn't tell whether she was blowing a guy or getting rescued from an Iraqi hospital.

(I thought it was interesting that relatives and loved ones of hers came to the show, all of them behaving as if they hadn't seen her video. I don't know how you have a family gathering when one of the people at the dinner table is famous for blowing a guy. What do you talk about?)

Like most people who get something without earning it, Paris didn't appreciate the experience. On Tuesday night, she visited all the writers in their offices, moving down the hall one by one. It's a tradition and an honor for a host to make those rounds, and they are happy to be there, jazzed to hear all the sketch ideas that are being written for them. I saw Paris come out of one writer's office holding her little dog, turn to her assistant with a slightly puzzled expression, and whine, "How many more of these do we have to do?" She resented having to meet these Emmy Award–winning writers. What an awful experience that must have been for her.

And then on Friday night, she did the unthinkable. The story that was going around was that she had locked herself in her dressing room, holding everything up for an hour and a half. On Friday night at *SNL*, there's no time to fuck around for even a minute. But Paris was offended to the marrow of her bones because one of the writers suggested that she do the cold open with Joey Buttafuoco, the man famous for banging an underage Amy Fisher, and then attempting to stand by his wife after young Amy *shot her in the face.* After his wife finally tossed him on his ass, Buttafuoco had cashed in on his celebrity from that scandal by fighting a female wrestler on *Celebrity Boxing.*

"How dare you compare me with someone like Joey Buttafuoco!" Paris complained with no irony whatsoever.

Lorne fixed the problem, Buttafuoco was dropped, and the show went on. But I'd never encountered anybody, no matter how big the star, who had the balls to say, "I won't go on." Luciano Pavarotti came and sang because he wanted to be on *SNL.* Hillary, Obama, Palin—none of them held up the show.

I will say one thing in Paris's favor, though. We rehearsed a sketch in which we were to play Donald and Melania Trump that included a full liplock—and yes, Ma, I know where her mouth has been, too. It didn't make the final cut, but we did do a sketch in which I kissed her on the cheek—she has absolutely the best skin of any human ever.

While no one else ever did anything that egregious, there were occasionally hosts who threw a wrench into the proceedings. In December 2000, Val Kilmer had the hosting duties. To me and my father, Val was the best Doc Holliday ever to be played on the big screen, which he did opposite Kurt Russell's Wyatt Earp in the 1993 film *Tombstone*. During Val's *SNL* monologue, we did a spoof of *It's a Wonderful Life* in which I played Clarence the angel, who walks Val around the studio, showing him how things would have gone if he had decided not to host.

Seconds before air, I was told by a stage manager that Val looked at one of the cue cards and said, "I'm not doing that one. I don't like it."

I was told, "Handle this."

By omitting the part he didn't want to do, Val left a gap that I had to bridge coherently off the top of my head. By then I'd been on the show for five years, so I'd gotten the hang of it, and I was able to walk us through the sketch without anyone being the wiser.

I remember one episode for the sheer beauty of the guests. Sting was the musical guest, and he had the best male physique I've ever seen. When I saw him during rehearsal, I thought, That yoga shit must work.

Elle McPherson hosted that same show. She was the first supermodel I'd seen up close. She was taller than I'd expected. Exquisite from top to bottom—God had obviously spent a lot of time on her. During rehearsal, she fell off the stage, and her enormous splendid frame plopped on the ground with a violence that would have put someone else in the hospital, but she got up and dusted herself off the way an incredibly well-constructed physical creature would.

When former cast member Phil Hartman came back to host, I tried to lay a compliment on him as he was entering the theater. He didn't respond. He just looked around and said, "Watch and learn, boys. Watch and learn." I don't know if he was serious, but it blew my shit away. I was in awe of him.

The host who spent the most time with me was Sylvester Stallone. We talked about the Everglades.

No, really, he was a big fan of the Everglades. It was early in my run at *SNL*, the beginning of my third season, and he left an autographed picture of himself in my office. He inscribed it, "Dear Darrell, Thank you for letting me bask in your talent." Nice guy.

In 2004, Robert De Niro hosted. As you might imagine, the proposed sketches were going to have at least *something* to do with the mob. Even my pal Eddie Galanek got in on the action. Drawing from a real conversation he'd overheard when he was working undercover with the Gambinos, Eddie wrote a sketch about two wiseguys having a conversation that went something like this:

"I gotta do what I gotta do."

"Then I gotta do what I gotta do."

"You're telling me that you gotta do what you gotta do because I gotta do what I gotta do? Don't tell me you gotta do what you gotta do because I gotta do what I gotta do, because I gotta do what I gotta do!"

Since it would have been odd for a script to be submitted with a member of the security staff's name on it, a writer named Lauren Pomerantz, who writes for Ellen DeGeneres now, and I submitted it under our own names. Everybody laughed at read-through, but another writer had also written a wiseguy sketch, and Lorne didn't want to do two, so Eddie's got cut.

But among the true highlights for me was the opportunity to meet two of my heroes. They weren't hosts, but if I hadn't been on *SNL* I probably wouldn't have met either one of them.

Eddie Murphy was the first impressionist I ever saw who was *always* funny. One day, he was doing *The Rosie O'Donnell Show*, which taped down the hall, and he strolled over to me and said, "I think your shit is freakish. Heh heh heh." He did his laugh. I was like, *Oh, my God. The distance between here and the closet in my childhood bedroom where I used to hide with a bottle of vodka cannot be measured.* I stumbled all over myself, trying to have a conversation with someone who inspired me so much. My ability to perform is limited to the stage; in private moments, I have no social skills whatsoever.

I met Richard Pryor near the end of his life when he was very sick with multiple sclerosis. I thought, Fuck it, you're my hero, and I have a chance to do this. I'm going to do it. I kissed him on the head.

CHAPTER THIRTEEN
Politics for Dummies

Washington, D.C.
2000s

O ne of the strangest things that came out of impersonating so many prominent politicians and news figures was that people thought I actually knew something about politics. Journalists were always asking my opinion about this, and my opinion about that, like I knew something. I'd tell them, "I'm a fucking clown." I'm not kidding when I tell you I have a copy of *Politics for Dummies* on my bookshelf at home. I was disheartened when I realized that to make what I thought was an informed judgment or decision, I would practically have to go to back to college and study economics, theology, history, and military strategy. I mean what, really, is a caucus? I still don't know.

I think they also wondered why I didn't hate anybody, why I never took sides. How do you hate a guy you're studying so closely? To do so would be to endorse the notion that there are people who come to Washington to *not* do their best, and I truly don't believe that, no matter what side of the aisle they're on.

There was a strange hybrid documentary about the 2000 presidential election that ran on PBS in the fall of 2002 that I was invited to participate in, but in character. So in between serious segments that addressed the procedures and the ultimate fairness of the election that put Bush 43 into the Oval Office, replete with policy wonks and analysts and whatnot, I'd pop up in the guise of Clinton or Gore or Cheney, and crack a joke. I'm not sure why producer Fred Silverman thought the combination of serious documentary and my humor was a good idea, but the end result baffled the reviewers.

In the fall of 2004, *SNL* put together a "Best of Darrell Hammond" episode, which made me the first cast member to have one of those while still on the show. (Let's just say, the show isn't available anywhere, and that's because it sucks, which was partly my fault.) In advance of the show's airing, Chris Matthews invited me on *Hardball*, which I agreed to do as long as he didn't ask me about politics. I guess Chris didn't get the message because when I got there, all he wanted to talk about was the midterm election and who I thought was going to win. He asked me about the Iraq War, for God's sake. I had no idea.

One network called me to comment on one of President Bush's State of the Union speeches. *Are you serious?*

I used to call in to Laura Ingraham's radio show a lot. I called in once as Oklahoma City bomber Timothy McVeigh. It was a great bit that she and I put together. She's talking to her sidekick, then all of a sudden the theme from *The Exorcist* begins to play. Laura is informed Timothy McVeigh, who had just lost his last appeal before his execution scheduled for a few days later, is on the line.

"So sorry, y'all, that music follows me wherever I go," I said.

We shoot the shit for a few minutes, and then she says, "You're going to go down as one of the worst murderers in history."

And McVeigh says, "In what sense?"

When the bit was over, I was played off by one of my old favorites, "Build Me Up, Buttercup."

Another time I went on her show with Chris and Kathleen Matthews. I played Chris, and he and Kathleen were, obviously, themselves. You can imagine the how crazy that got, with *two* Chris Matthewses screaming emphatically at the same time.

Laura and I hung out together a fair amount, and she was instrumental in the first time I did Rumsfeld. She was going with me to Rosie O'Donnell's studio, and we came up with a bit in the car together.

In character as Donald Rumsfeld: "Al Quaeda says they're a spiritual people who want to find God by dying in a war with the West. To hell with it, let's help 'em out!"

I was totally impressed by Tim Russert, whom I met backstage at an event I did for NBC president Jeff Zucker at the St. Regis in New York, where there is a well-used ballroom on the twenty-third floor for banquets and such. Russert and I chatted, but I didn't want the conversation to go too far because I worried that pretty soon he was going to ask me about the budget or, Who's your financial planner?—shit I didn't know anything about.

Eddie Galanek sometimes did security for Jim Cramer, the CNBC money guy, when he was on the road, so Eddie got this idea that Cramer and I should meet. With Cramer's outlandish gesturing, Eddie thought he'd be a great character for me to do on the show. One afternoon, Eddie took me out to Englewood

Cliffs, New Jersey, to the CNBC studios. When we got to Cramer's office, he was going on about this one company, explaining the product they made, where they were going with it, the profit they were going to make. Then he said, "I'll be right back, I just gotta go grab something." As he walked away, I turned to Eddie and said, "He doesn't know I'm a moron about this kind of stuff?"

Maria Bartiromo showed me around the stock exchange one day and tried to explain how it all worked. Finally, I had to interrupt her. "I don't understand any of this. There are *numbers* everywhere."

But it was apparent that some of the media had *really* confused me for the people I played on television. Deborah Norville's people called me feverishly for two weeks to get me on her show. When I showed up in the studio, she walked right by me. She had no idea who I was out of costume. In the same vein, I was invited to Larry King's show—I'd met him and his wife at an event in D.C.— but on the air he called me Darrell Hayman.

Regardless of my patent ignorance, a lot of the power players I impersonated on television invited me to perform for them privately.

On the back of my Ted Koppel impression on *SNL*, I was invited to an event honoring the celebrated newsman at the Museum of Broadcasting. They had me come dressed as Koppel, with the big orange wig and the fake nose. I stood outside the room where the party was being held while Sam Donaldson stood at the podium, drink in hand, giving a slightly moist appreciation of Ted Koppel. I overheard him say something along the lines of, "Then we chartered a plane and we went to St. Louie. I said, 'Ted, to St. Louie? You test the elasticity of my credulity.'"

Then it was my turn. Everyone turned to look at me as I walked into the room, and I was terrified to see the great man himself. I turned to him and, in his voice, I said, "Are you mad at me?"

He looked at me and said, "No, I'm not mad at you. Give me the microphone."

I handed him the mic, and he turned to Roone Arledge, who was the president of ABC, and said, "Roone Arledge, you cheap bastard, if you paid me a living wage I could afford a decent rug like this guy's got on."

The audience laughed and applauded, and I was allowed to leave.

At a party at the Waldorf, I found myself on a panel near former president George H. W. Bush. I told a joke I told at many business conventions: "President Bush and President Clinton have become good friends, and President Bush has only served one term, so he could run again. Why not name Bill Clinton as his V.P.? I'd like to be a fly on the wall for that conversation.

"Clinton says, 'Let me get this straight, you want me to be your vice presidential candidate? That would really piss off Hillary.'

[Pause for effect.]

" 'I'll do it.' "

The Waldorf exploded. And I was delighted to see I'd made another president of the United States laugh at my jokes. I chatted with him for a few minutes beforehand at a cocktail reception, and he told me how much he loved Dana Carvey's impression of him. But I was always afraid the president said to the Secret Service later: "That guy's a dick." What if you got called a dick by a president? You'd take it to heart, wouldn't you?

One Washington insider told me, "You know, you're going to have to take one side or the other if you're going be a player in this town." I was like, "I don't *want* to be a player."

Thanks to Bob Barnett, who was both Clinton and Cheney's lawyer, I got to have the extraordinary experience of being alone in a room with President Clinton. I was doing *The Tonight Show*, and Clinton was giving a speech to the California Bar Association at the Beverly Hills Hotel, where I was staying. When I found out he was going to be there, I called Bob and told him I'd love to shake the president's hand, if that were at all possible. The next thing I know, I got a call in my room: "The president will see you." So I went downstairs to a meeting room where he was cooling his heels until his speech. He greeted me like we were old friends. At first there were a couple of Secret Servicemen, then they excused themselves and it was just us. There was a little table with food and soft drinks (nothing says you've arrived at the inner sanctum of power like light refreshments). At the time, the former president was working in conjunction with the first President Bush on disaster relief efforts in New Orleans after Hurricane Katrina had destroyed the city.

I heard my voice say, "You like him, don't you?" I couldn't believe how cheeky that was of me. Like this was any of my business.

Then I heard his voice say, "I do. The only difference between us was domestic policy."

Did anyone else hear that?

Then he went out and spoke for forty-seven minutes and got a standing ovation. Grammatically perfect, vivid, lucid, powerful prose. He mentioned me twice.

Thanks to my Cheney impression on *SNL*, starting with the 2000 election, I was invited to George W.

Bush's second inauguration in January 2005. I had access to the White House if I wanted to go hang out. During the main event, Cheney rose up out of the floor in the middle of Madison Square Garden, his hands in his pockets, surrounded by 35,000 revelers. It was the greatest theater I'd ever seen, Cheney with that lopsided grin on his face looking out at the world like a conquering hero. It was an unimaginable display of power.

Afterward, Cheney had me up to his box. He shook my hand warmly as he always did. He introduced me to a man whose eyes seem to suggest he knew all of the secrets to the universe. The man looked at me the way the workers at Sea World would look at the fish. Cheney said, "This is Darrell. He's the guy who does all the damage on *SNL*. Heh heh heh." And the guy just looked at me.

Cheney patted me on the shoulder as if to say, "He's not so bad. He's just doing his job."

As part of the inaugural festivities, I was invited to perform for the troops at an event hosted by Kelsey Grammer at the MCI Center in Washington, D.C. Bush 43 gave me a tremendous bear hug and said, "You lost quite a bit of tonnage." I have to say, it's different getting a personal comment from the president of the United States than from, say, a bus driver. *What are* *you suggesting, Mr. President?* But I thought it was a funny line, and I laughed.

During my performance, I said, "Mr. President, not many people get a chance to talk to you. I'd like to make use of this opportunity, because I know I could either crack wise or try to make a difference. So I would like to ask you, Mr. President, is there anything you can do about Brad and Jen?"

He laughed hard. Once again, the thrill of hearing a president laugh at something I said. But it should be noted that joke was written by Beth Armogida, who writes for Steve Martin and the Oscar show, and who has saved my ass on more than one occasion by sending me last-second one-liners.

Cheney was the best laugher of all. He didn't have Clinton's famous charm, but I'd say shit and he'd crack up. He was always very nice to me. I ended up performing for him at his Christmas parties, at White House functions, and at various private events.

Once, the vice president summoned me to the Greenbrier Resort in White Sulphur Springs, West Virginia, about a five-hour drive from Washington, D.C.—one of the most fascinating, unsettling gigs I ever did.

The Greenbrier was famous as a congressional hideaway, filled with subterranean bunkers built in the 1960s during the Cold War. The vice president's people had sent a car to pick me up at the airport, and I was hustled through a maze of underground rooms (the vice president was still taking precautions in the wake of 9/11) until I found Cheney and about forty of his buddies—including Trent Lott, Tom DeLay, Dennis Hastert—eating and drinking and laughing.

Once again, it wasn't so much Darrell Hammond who was invited but Bill Clinton, so I showed up in character, ready to poke some fun at the gathered Republican power brokers. I started out with a joke about Cheney's pacemaker—that every time he sneezes, garage doors start to open all over the neighborhood.

At one point, I looked out across the room and thought, Man, it doesn't get any weirder than this. In fact, it was weird *squared*. After I'd zinged the key players, I spent the rest of the evening hanging out with these guys. There didn't seem to be any reason for the gathering; it was just a bunch of pals having a good time. Everybody wanted to pose for a photo with "Bill Clinton." There was one great shot of me reaching out to shake Cheney's hand, the vice president recoiling in mock horror.

When I attended an event at the vice president's residence in 2006, where I would be simply myself, Lorne took the precaution of sending his fantastic assistant Lindsay Shookus, who is as good a conversationalist as has ever lived, to accompany me. When I was introduced to Chief Justice Roberts, it became immediately apparent that I didn't know how to talk to a guy who was fourteen or fifteen times smarter than me. *You don't have to keep talking to me. You're a very nice guy, Chief Justice Roberts, but you and I both know that pretty soon you're gonna get bored with this. You're going to want to talk about torts and meaningful shit, books you've read about the galaxy, and I won't know anything. And that will be bad. I'll be embarrassed.* He was very nice, but I was relieved when Lindsay took over.

Brit Hume was there with his wife. Brit was very sweet to me, but his wife walked up to me and said, "Cheap shot." I didn't know what she was talking about, but I assume it was because of the bit I'd done as him during the third episode of that season. Fortunately, Lindsay went over and talked to her, and they were all smiles after that.

But I didn't like that part of it. I really liked these guys—all of them. Hell, I was impressed that they would even talk to me. By the time someone introduced me to Scooter Libby, Cheney's chief of staff, and a couple of four-star generals, I thought, I've come far through the looking glass.

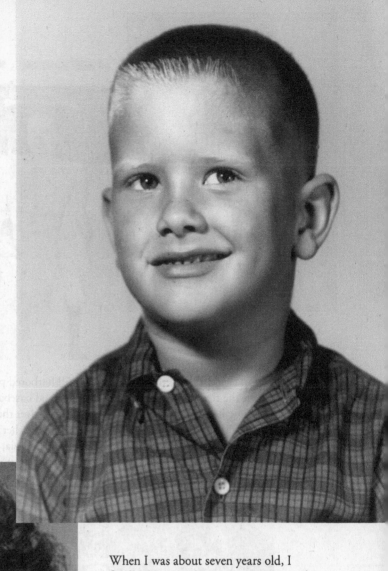

When I was about seven years old, I finally discovered a way to connect with my mother (*below*). I noticed that if I could get her talking about certain people in the neighborhood, she would become enraptured doing impressions of them. Doing my best to copy what she did, I learned to do voices too. If for nothing else, she seemed to love me for that.

Photos courtesy of the author

My father, Max, was sent to Germany an eighteen-year-old second lieutenant, the youngest commissioned officer in the history of the U.S. military at the time, or so he was told. He came home with a lot of medals and a tortured soul.

He still harbored plans to play professional baseball and go to law school after the war, but then he and my mother got married in Sylvester, Georgia, in 1947, so he had to work to support his wife.

Three years later, my Dad (*center*) was called back for duty in Korea. A week after that he shipped.

Photos courtesy of the author

Between the ages of twelve and fourteen, if I wasn't playing baseball, I was dreaming about it. Every day, the second I woke up, I wanted a bat in my hand. That cocky blond kid on the top left is me during the All-Star game in the Babe Ruth League. Sometimes before going out to do an *SNL* sketch, I'd think about the two doubles I hit off of C. P. Yarborough (*2nd from right, rear*) in one night. *Photo courtesy of the author*

Working at the University of Florida radio station on a series about suicide prevention. I interviewed students who had attempted to kill themselves as well as health professionals commenting on why people do it. A congressman was going to hold a press conference about it, but it was cancelled at the last minute when the spots were deemed too disturbing to air. *Photo courtesy of Michael Kooren*

In the 1980s, I sometimes hit the road with a kid named Billy Gardell, who now stars in the sitcom *Mike & Molly*. He was only eighteen then, but he already wrote brilliant stuff. *Photo courtesy of the author*

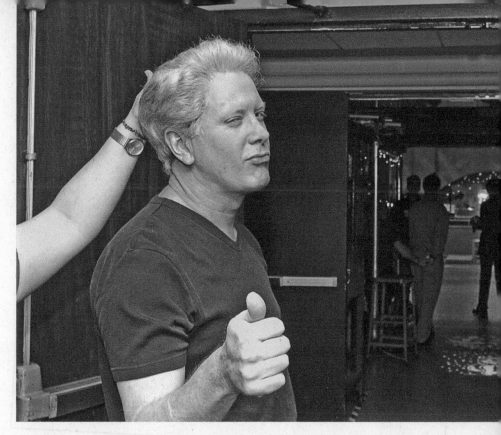

Outside the doors to Studio 8H, getting ready for rehearsal. The first time I did Clinton biting his bottom lip and giving the thumbs-up on air, the audience loved it. After that, anytime I got a Clinton script, it would include a note, "Does the thumb and lip thing."   *Credit: Mary Ellen Matthews / Courtesy of Broadway Video Enterprises and NBC Studios, Inc.*

Monica was a great kisser.
Photo courtesy of the author

To Darrell Hammond with thanks for filling in the last few months — that's our secret! Best wishes —
Hillary Rodham Clinton

Mrs. Clinton sent this photo after I appeared as the president's clone at the 1997 White House Correspondents Dinner. The inscription reads: "To Darrell Hammond with thanks for filling in the last few months—that's our secret!" *White House photo*

I wore the white wig to play the forty-second president more than eighty times while I was at *SNL*. Although I played more than 100 different characters over the years, Clinton would largely come to define my time on the show. *Credit: Courtesy of Broadway Video Enterprises and NBC Studios, Inc.*

DUPDATE|WEEKENDUPDAT

As Phil Donahue on "Weekend Update" with Tina Fey during my first season. I had done Donahue in Spanish when I auditioned for *SNL*, and it became one of my first successful impressions on the show.   *Credit: Courtesy of Broadway Video Enterprises and NBC Studios, Inc.*

On the back of my Ted Koppel impression on *SNL*, I was invited to an event honoring the celebrated newsman at the Museum of Broadcasting. When I took the stage dressed as him, he turned to ABC president Roone Arledge and said, "Roone Arledge, you cheap bastard, if you paid me a living wage I could afford a decent rug like this guy's got on." It got a huge laugh. Sam Donaldson is making the rabbit ears.

Photo courtesy of the author

As *Hardball*'s Chris Matthews. I was once on Laura Ingraham's radio show in character as Chris Matthews along with the real Chris Matthews. You can imagine how crazy that got, with two Chris Matthewses screaming emphatically at the same time. *Credit: Dana Edelson / Courtesy of Broadway Video Enterprises and NBC Studios, Inc.*

As presidential candidate Al Gore alongside Will Ferrell as George W. Bush, the Supreme Court in the background, when Gore finally conceded the 2000 election. When I studied Gore for the part, I realized that in each debate he spoke in a completely different way. I decided he'd probably had three different voice coaches. My impression combined the three. A reporter for *U.S. News & World Report* said I lost the election for Gore, which made me feel terrible. *Credit: Courtesy of Broadway Video Enterprises and NBC Studios, Inc.*

When I did the first Bush White House Correspondents Dinner in 2001, I gave the new president a baseball glove. Fortified by a bottle of wine, I proceeded to do a lot of Gore material. I turned to Mr. Bush and asked, "Can I do some more of him? I worked a year to learn the damn guy, and then you beat him."

Photo courtesy of the author

THE WHITE HOUSE
WASHINGTON

May 3, 2001

Mr. Darrell Hammond
Saturday Night Live
30 Rockefeller Plaza
Room 1720-E1
New York, New York 10112

Dear Darrell,

Thanks very much for the glove and ball. I look forward to playing catch with you on the South Lawn some time.

Your remarks at the White House Correspondents' Dinner were very entertaining. Laura and I enjoyed your performance.

Sincerely,

George W. Bush

Thanks for making us laugh

As Dick Cheney in October 2001, the day after it was announced that anthrax had been found at 30 Rock. In the sketch, Cheney reveals his secret location in Afghanistan and his one-man mission to hunt down Osama bin Laden with the help of his "bionic ticker that monitors my heart-beat, gives me night vision, and makes me completely invisible to radar." The real veep told Rush Limbaugh he got a kick out of the bit. *Credit: Courtesy of Broadway Video Enterprises and NBC Studios, Inc.*

With Vice President and Mrs. Cheney. The Vice President invited me to his private box during the festivities at the second Bush inauguration. He intro-duced me to one of his pals: "This is Darrell. He's the guy who does all the damage on *SNL*. Heh heh heh." *Photo courtesy of the author*

As Senator John McCain, with Fred Armisen as then-Senator Barack Obama. McCain had visible war wounds, and part of imitating him would be to imitate his physical limitations, which were the result of torture, and I didn't want to make fun of him. But my father, not knowing what this would entail, desperately wanted me to play McCain, whom he thought was a great man. In the months before my father died, he said, "That's my final wish." I was never able to pull it off well.

Credit: Mary Ellen Matthews / Courtesy of Broadway Video Enterprises and NBC Studios, Inc.

Lorne Michaels with Sarah Palin watching Tina Fey in costume as Palin on the monitor. When Palin did a cameo on the show, she had more Secret Service protection than any other politician I'd ever encountered. I was mortified when she decided to use the bathroom outside my dressing room—what if I sprinkled when I tinkled? *Credit: Mary Ellen Matthews / Courtesy of Broadway Video Enterprises and NBC Studios, Inc.*

About a month before my father passed away in 2007, I thought it might give him a thrill at the end of his life to see his son hit some baseballs. I got in touch with my old high school friend Wayne Tyson and asked him if he'd pitch to me. He agreed, and offered up the field at Palm Bay High School, where he was coach. I hadn't swung a bat in thirty years, but muscle memory kicked in. We didn't talk about it, but I knew we were all thinking about that 297 sign at Wells Park and hitting one over the wall. *Photos courtesy of Wayne Tyson*

My friend Eddie often accompanied me when I flew down to Florida to see my father in his last year. Eddie joined me and Wayne on the baseball diamond that day. That's me in the outfield.

My father on the bench in the dugout, watching the proceedings as carefully as he ever had. His right ear, out of sight here, had recently been removed in a futile effort to combat his cancer. When I hit a high fly ball that nearly went over the fence, he struggled to his feet and paced it off, just like the old days. Later, he said, "It woulda gone out."

In a sketch as Donald Trump, with Paris Hilton as Trump's wife, Melania. When I kissed her cheek during the bit, I discovered that she has absolutely the best skin of any human ever. *Credit: Mary Ellen Matthews / Courtesy of Broadway Video Enterprises and NBC Studios, Inc.*

Back at *SNL* in May 2011 for a cameo as Donald Trump in a debate sketch that also featured host Tina Fey as Sarah Palin. That's the fabulous Jodi Mancuso getting me ready for the Trump wig. Tina's Palin wig is draping the counter in front of me. Note all the head forms on the shelf behind me; each cast member has his or her own.

Photo courtesy of Elizabeth Stein

When Donald Trump hosted the show, he dove in with tremendous enthusiasm. He asked more questions and spent more time with the cameraman, the
lighting people, and the costume people than just about any other host I'd ever
encountered. But he's a lot taller than I am, so in our matching dark suits and
purple ties, I ended up looking like a Mini-Me version of the dude instead of
his double. *Credit: Mary Ellen Matthews / Courtesy of Broadway Video Enterprises and NBC
Studios, Inc.*

After three months of in-patient rehab, I'm back on stage in spring 2011 at Caroline's, where Marci Klein had discovered me for *SNL* sixteen years earlier.   *Credit: Ayala Gazit*

As Truman Capote in the one-man play *Tru* at the Bay Street Theater in Sag Harbor, New York, June 2011.
Credit: Jerry Lamonica

CHAPTER FOURTEEN

My Welcome Outstayed Me

New York City and Melbourne, Florida
2006–2009

A t the beginning of my twelfth season in 2006, I was in Lorne's corner office, with its spectacular views of New York out the window, for the first glorious pitch meeting of the year. I was at *SNL* again. *Ah, man, I must be the luckiest motherfucker who ever lived.*

But that year, as I sat in that office while everyone introduced themselves to that week's host, Dane Cook, another thought crept into my brain: *I should not be here.*

I called my manager, Bernie Brillstein, and said, "Isn't it time? Isn't eleven seasons enough?"

"No, that's your place," Bernie said. "You belong at *SNL*."

Kenny Aymong, the supervising producer and one of Lorne's top people, told me, "This is your home."

"I know, but it's time."

"Not this year," everyone seemed to be saying.

So I stayed. For Lorne, for Marci, for Bernie, for the show.

It was awe-inspiring to see people like Tina Fey, Steve Martin, John Cleese, asking Lorne about a joke. Jay Leno would call Lorne to ask about a joke. How could I let him down?

Marci is among the more interesting human beings I've ever met in my life. She used to walk up to me and say, "Who's your rabbi?" And I'd say, "Well, you are, boss." And she'd start laughing, toss her hair, and walk away. I was once chatting with her at a party, and JFK Jr. came up behind her and tapped her on the shoulder. She turned around to him and said, "I am fucking talking, okay?" (I'm not gay, but if I was, damn . . . he was a good-looking man.) But that's how loyal and how powerful she is at the same time. How could I let her down?

And yet, by then Clinton had been out of office six years, Gore had been off the political scene just as long, Will Ferrell had left the show, so his Alex Trebek no longer hosted Sean Connery on *Celebrity Jeopardy!*; in those final years I was left mostly with an occasional

Dick Cheney and John McCain in a cold open or in a "Weekend Update" segment.

There was one show that stood out, at least to the press, and that was when I did Fred Thompson in October 2007. Thompson, who had been a two-term U.S. senator from Tennessee, was most famous for his role as the district attorney on *Law & Order,* which he took up when he gave up public office. But it seemed his popularity on the show convinced him that he had a shot at the presidency, the real one, so he tossed his hat into the Republican ring, although not very convincingly. He announced his campaign on *The Tonight Show* in September in an effort CNN called "lackluster" and the *Chicago Tribune* called "awkward." I heard that Thompson's people were pleased to learn that I was going to be doing him on the show, although I imagine that changed when they learned we were making some fun of his apparent lack of enthusiasm—"How badly do I want to be your president? On a scale of one to ten, I'm about a six." The *New York Times* said my portrayal was "potentially devastating" to his campaign. I find that hard to believe, but then Thompson withdrew three months later. I was slated to do him one more time, but I couldn't get the voice right—by then I had my stalker and I was scared out of my wits.

Which no doubt contributed to the fact that in those final seasons, I was on air so infrequently that people started to ask me if I was still on the show.

I had several years of post-Hazelden sobriety under my belt, but my shrinks still had me on major meds. I did an episode of *Law & Order: Criminal Intent* in 2005 on Lamictal, a mood stabilizer used to treat bipolar disorder, and Zyprexa, an antipsychotic. On just one of those, whatever you are as a human being stops. You simply aren't there. You don't feel happy, you don't feel sad, you feel nothing. And I had to do scenes with Vincent D'Onofrio, a world-class actor. How the fuck was I supposed to do that?

I even made my Broadway debut, playing Vice Principal Douglas Panch in the Tony Award–winning musical comedy *The 25th Annual Putnam County Spelling Bee*, at Circle in the Square for six weeks during the summer of 2007. Don't worry, I didn't sing. Although I wish I could have sung "My Unfortunate Erection"—I think I could have brought a lot of heartfelt emotion to that number. Instead, my character administered the spelling bee of the title and said things like, "That is correct." But don't think it wasn't challenging; I dare you to pronounce *acouchi, boanthropy, mohel, corzya,* and *Ilspile* in front of hundreds of strangers without getting tripped up.

I did Christopher Durang's comedy *Beyond Therapy* in 2008 at the Bay Street Theatre in Sag Harbor. Working with such outstanding actors as Tony-nominee Kate Burton and Tony-winner Katie Finneran, I knew the only way I could hold my own was to quit taking my meds for the run of the show, so I did, and it worked.

I had a recurring role as a creepy hit man called The Deacon on Glenn Close's show *Damages* during my final season on *SNL*. Seth Meyers, who became head writer when Tina Fey left the show in 2006 was a big fan of the show, and he was very complimentary. I think it gladdened him to see me get great reviews when I wasn't doing much on *SNL*. Aside from being a great writer, Seth was another one of those guys who reached out to help me when I was down. I won't forget it.

In 2005 my mother, a lifelong smoker, was diagnosed with lung cancer. Her doctors wanted to do chemo, but she said, "For what?" She was going to go to her death like Ted Bundy to the electric chair.

I flew down to see her shortly before she died in February 2006.

I hadn't spoken to or seen her in several years, although I'd heard that she and my father never missed an episode of *SNL*, even when they were angry with me for confronting them about what happened in our house when I was growing up.

I brought Eddie Galanek with me. After years of looking after me when my stalkers were around, Eddie had become a good friend. But it was also true that back in 2001, my father had threatened to shoot me. Just a few weeks before 9/11, I made one of many appearances on the Howard Stern radio show. On this occasion, we got onto the subject of my addiction problems, which led to a conversation about childhood abuse. I didn't get into the details with Howard, but enough of the interview made its way into the Florida papers, and my father was pissed enough to threaten me. I knew he was now enfeebled, but I wasn't taking any chances. Fear will not be denied.

At the hospital, I approached my mother's bed, where she lay like a queen who's used to running the show. She grabbed my hand and said, "You were always my buddy, right?" She always called me "Buddy" when I was a kid. "You were always my favorite. You know that, don't you, Buddy?" Her Southern accent was perfect.

She looked at me as if to say, *I win, right?*

Eddie and I were barely back in New York when we got word that she'd been moved to hospice, so we turned around and headed back to Florida. Two weeks after our first visit, she was gone.

My father took her death with his customary stoicism. We were gathered around her bed, talking quietly. One minute my mother was breathing audibly, and the next she was silent.

My father looked at her and said, "And this is what happens."

My parents, both of them, were immensely popular people in their community. The church was packed with well-wishers for my mother's funeral.

After it was over, people came up to me, clasped my hands, and said, "Your mother was such a wonderful person."

"She helped us out so much."

"I remember when your mother brought a casserole when my husband was sick."

I went into the bathroom and stuck my tongue out at the mirror, the scar winking at me. I pushed up the left sleeve of my shirt and ran the fingers of my right hand over the ladder of scars running up and down my left arm, each rung a different shade of pink or red, depending on how recently the wound had been inflicted. I understood then that people are capable of leading double lives.

I also realized that I'd been holding on to a fantasy that I'd one day be able to reclaim my childhood. Here I was, fifty years old, and I was still hoping for a do-over. At long last, I knew that wasn't going to happen.

My father was already sick with advanced melanoma when my mother died. A soldier to the end, he wanted to fight it. He went through chemo and multiple surgeries, all of them horrendous. They cut out chunks of flesh behind his ear, trying to remove the tumor. Finally they cut his ear *off*. And every time, he said, "Son, I'm gonna beat this sombitch."

After years of estrangement, my father and I talked a lot in the months leading up to his death. I flew down to Florida with Eddie from time to time to see him. Even after it was clear my father posed no threat, Eddie kept coming with me to help out, and the three of us would sit together, watching old war movies and John Wayne westerns, or going through my dad's high school yearbooks. On one occasion, my dad dug out his war medals and explained what each one was for.

Over those last months of my father's life, Eddie often cooked for my father, and the two men grew quite close. Eddie had seen his share of action when he was on the police force, and he'd spent months down at Ground Zero looking for remains after 9/11, so they shared the experience of having lived through man-made horror. During one medical visit, the doctor talked to both Eddie and me about "our" father; neither one of us corrected him.

My father talked a lot about what he was going to leave me in his will. Amazingly, I found myself getting sentimental about it. Yeah, it *would* mean something to me to have his television, his suits, his watch. He didn't have any money, and I certainly didn't need any, but it was very emotional to hear him talk about the legacy he wanted to give me. Even more moving was how he kept telling me he was sorry he hadn't been around for me when I was growing up. I didn't know if he was referring to how much he was on the road for his job, or something more, and I didn't ask, but it meant a tremendous amount to me to hear him say it.

And I tried to be a son, to do everything I could to provide for him—a private nurse, whatever he needed. It was my last-ditch effort to have some scrap of childhood, a trace of a parent who cared about me.

A few weeks before he died, when he was filled with painkillers and forgetting things, my father told me that he left the glass door to the patio open, and a cottonmouth came into the house. We were always killing cottonmouths in the yard when I was growing up.

My father grew tomatoes and basil on our patio. There was often an enormous black snake out there, coiled up, sunning. My father said, "That's old Blacky," although the way he said it, it sounded more like "Black-ay." The snake ate rats and other snakes. You couldn't fuck with my father's tomatoes because of Old Blacky. But that snake never bothered us.

However, more than once my father got me out of bed to come out in the yard to kill a water moccasin. FYI, they don't kill that easily. They really need to be shot, but he didn't have any ammo. He'd be out there in his coaching shorts and a T-shirt and flip-flops, our husky howling nearby. He'd have the snake pinned with a shovel. "Darrell, help me kill this sombitch."

He used to take me Sunday fishing, him and his buddies, to a wonderful place to catch bass. To get there, we passed through a part of the stream where it narrowed to about ten feet, maybe six feet. And on each side were stumps, and on each stump, there was always a moccasin. Every Sunday, my father executed them with an enormous .44 Smith and Wesson. I don't care what kind of creatures they were, it wasn't fun to see them executed. The sound wasn't just deafening, it was injurious, as the snakes evaporated into a fine red mist. And then we'd go fish.

One time, I was with my father driving from Tampa to Melbourne on Route 192, and my father saw the tail of a rattlesnake exiting the highway into the bushes. He'd lost two dogs as a kid, one to a rattlesnake, one to a cottonmouth, so whenever he saw a snake, it was

Onward Christian Soldiers. He stopped the car, got out, retrieved some empty bottles from the trunk, then carried his arsenal over to the guardrail, where he perched himself while he chucked his weapons at the snake.

Now, eighty-five years old and in a narcotic delirium, he claimed he had shot that snake that slithered into the house. There were no bullet holes in the floor, but we didn't question him. Can you imagine the amount of medication you'd have to be on when they cut your ear off? You'd probably think you'd gone Rambo on some reptiles yourself.

During the afternoon of Saturday, November 3, 2007—the day that Barack Obama was to appear on *SNL*—a nurse from the hospice where my father had finally been transferred when he needed full-time care called. "Your father may not have more than a day or two left. You need to get here."

But my dad was insistent that I do *SNL* that night because of Obama. It was an honor, he said, to appear with the man who might become our next president. My father had the hospice nurse take him off his morphine drip because he was afraid he would fall asleep and die before I could get there.

On Sunday, Eddie and I flew down to Florida. When we approached what was to be his deathbed, he said, "Boy, you did so good on the show last night."

He had his war medals in their plastic cases by his side. My sister must have brought them to him, although perhaps he'd taken them with him when he was admitted. Once again, he took us through each one, explaining how he got them. I figured it was his way of saying, "What I really did while I was on earth was fight for my country against Hitler." It was the best he could do to explain himself. It had meant a lot to him when, a few months earlier, Vice President Cheney had written him a personal note of thanks.

And one last time, he said, "Darrell, I just want you to know how much I love you. I'm sorry for anything that I might have ever done that upset you, that I wasn't there for you when you needed me."

And then he turned to Eddie. "Eddie, I'll see you somewhere on down the road, son."

Then to the nurse, "Now give me my fucking morphine." I can't imagine how much pain he'd endured to make this good-bye possible, but it was clear from his face he'd had enough. It wasn't long before the drugs did their trick.

Eddie, my sister, and I stayed in the room, my father unconscious but hanging on. We didn't want him to be alone when he made his final journey.

After a time, a nurse came in and said, "Sometimes they won't let go if there are loved ones in the room."

It had been maybe three or four hours since we'd arrived. The three of us stepped out of my father's room
and down the hall to a small patio. Eddie said something wickedly funny that had all three of us howling
with laughter when the nurse came to get us a few minutes later. She stood silently behind us. We turned to
look at her.

She said, "He's gone."

It was that fast.

At the funeral, I thought about this man who had
lived in a bottle of gin since he was a kid because he'd
never gotten over killing a lot of people in service of
his country. For fifty years, he struggled with it. Was
it right to have killed so many people? At the end of
his life, he was able to say, Yes, it was. Someone had to
stop what he called verifiable evil. He wasn't referring
to the men who end up in the penitentiary because
they caught their wife cheating, got drunk, lost their
temper, and clocked a guy with a blunt object. Nazis
are different. Maybe Korea wasn't as cut and dry, but
at the time the Communist threat seemed like something that needed to be stopped, and war was the only
way to do it.

Eddie was incredibly helpful in making the funeral
arrangements when the rest of us were in no shape to

do so. He even gave the eulogy, which went something like this:

I know most of you have known Max longer than I've known him. I met him through Darrell, and we've gotten very close. I'm sure all of you have had an experience with Max where maybe he had a little too much to drink and he said something harsh to you. But there was a boy who wanted to play baseball and be a lawyer. He wasn't given that choice. His life in the military, doing what he had to do, it changed who he was. But the boy always lived inside of him. Every one of you knows how much he loved baseball. Every one of you knows Max, the other guy, the soft guy, the baseball guy. He didn't have the choice but to become the hard man, but when he came back around, as you all know him now, he was a sweet guy. I know I'm blessed to have met him. He did a lot for this country, and he's a good man.

I know in my heart that Eddie didn't give that speech for those old folks in attendance, or even for my father—he gave it for me.

When the two soldiers from the U.S. Army played "Taps," and they brought my father's folded flag out,

I felt so much pride, and so much sorrow that I hadn't understood that I had grown up in a soldier's house. He'd fought evil, and it had fucked him up.

I convulsed with tears and thought I'd never stop. Big, tough Eddie, a man who'd been to hell and back himself, put his arms around me.

As we left the church, I remembered a time when I was about twelve years old, and I came upon my father sitting in our blue vinyl chair, mumbling. In his lap, he held a Luger he had gotten during the war. It wasn't loaded; he wasn't going to shoot himself or anyone else. He was just thinking about the German officer he'd gotten it from.

"He looked at me, and he thought he was right," he said. My father was haunted by that man's eyes. He paused, and then he said, "I took his gun."

He didn't say he killed the man, but how do you imagine he got the gun away from him?

I said something like, "Why would you take the gun?" I was a kid, I didn't understand humiliating an officer.

"I wanted him to remember what he saw in my eyes, and that if I ever see him in the afterlife, I'll kill him again."

I wondered if my father had found him yet.

My parents were both cremated, their ashes buried in the courtyard beside First United Methodist, the church where I spent my earliest years frightened out of my wits.

A few months later, I was getting ready to perform at a convention in Orlando when the producer of the show came up to me and said, "We have a mind reader here. Would you like him to read your mind?"

Sure, why not?

The mind reader came over to me and asked me to write down a couple of things. Then he took the paper and tore it up, keeping the bits and pieces balled up in his hand.

"Who would you like to know about that's deceased?"

"My father."

He held the paper, and concentrated on it for maybe fifteen seconds.

"He's wearing a baseball cap, and he's sorry."

As I had when my sponsee Dean killed himself two decades earlier, I. Freaked. The. Fuck. Out.

The day after my father passed away, the Writers Guild of America went on strike. More than ten thousand writers were demanding their fair share of DVD and Internet residuals. I got back to New York, only to discover there would be no new episodes of *SNL* until it was resolved four months later.

Not knowing how long the hiatus would last, the writers and cast decided to take the show on the road, or at least downtown, to the basement theater of the Upright Comedy Brigade. Since the show would not be televised, and no new material could be written for it, we used previously written sketches that hadn't made air for one reason or another. The end result was a slightly raunchier show than we could have ever done on network television. Proceeds from ticket sales to the live show went to the production staff, which had been laid off for the duration of the strike. We did our own makeup, gathered our own props, the writers held the cue cards, and Gena Rositano, the stage manager, walked around wearing a headset that wasn't plugged in. Amy Poehler, who is a founding member of UCB, invited Michael Cera, who had recently starred in the movie *Superbad*, to host. The writers cleverly cobbled together a monologue for him from what past hosts, including Snoop Dogg and Paris Hilton, had performed. At one point, I went onstage with a hat filled with pieces of paper on which I had scribbled celebrity names. I pulled them out at random and did impressions. The show was a huge success with the 150 or so people who came to see it.

Neither NBC nor Lorne had anything to do with this renegade production, but he was in attendance. And when it was over, Lorne, whose birthday it was,

bear-hugged me, which I took as his way of express-
ing condolences for my father. In deference to such an
intimate moment, everyone else looked away like the
Queen was taking a leak.

The season resumed in February, with Tina Fey
as host. I was barely in the remaining episodes of the
season, which was fine with me. My heart wasn't in it
anymore.

I believe the greatest mistake I ever made was going
back to *SNL* for the final year. Why get in the boxing
ring with Mike Tyson when you're not in the mood?
That's how hard it was.

About a week before that season was to begin, I went
back to the ER. I said to the nurse, "I think somebody
poisoned me."

"Who poisoned you?"

"I don't know, man, but I feel like I'm poisoned."

They took my blood pressure, and it was about 180.

"You've got a heart problem."

They ran tests on me all night long.

In the end they said, "There is nothing wrong with
your heart. We think you're afraid."

In the closing months of 2008, everything was about
Tina Fey's outstanding impression of Sarah Palin,

which was fine by me because it took the focus off my McCain. I didn't want to play him. I didn't want to play a soldier after watching my father dying a soldier's death. McCain had visible war wounds, and part of imitating him would be to imitate his physical limitations, which were the result of torture and I didn't want to make fun of him.

But my father, not knowing what this would entail, desperately wanted me to play McCain, whom he thought was a great man. In the months before my father died, he said, "That's my final wish."

What could I do? I played McCain several times that last season, and it was awful. I never could do the impression all the way. It felt like hell.

Senator McCain himself made a few appearances on the show, and one day I ran into him in the hall while he was talking to Jeff Zucker. Much to my surprise, McCain hugged me. Then he turned to Zucker and said, "I just want to say, Darrell Hammond does the best impression of me there is." I thought, How fucking gracious is this guy? He knew my work, he had to know it wasn't any good, but he complimented me to the big boss.

Everybody at *SNL* knew my McCain was off, but nobody came to me and said, "Hey, what's wrong with you? How come you can't kick the ball between the uprights on this one? What's going on?" I guess they assumed I would correct it eventually.

In the final months of my fourteen-year tenure at *SNL*, I was circling the drain personally. I'd pretty thoroughly relapsed, I was on medication that had put thirty pounds on me, and I was miserable. I got it into my head that all of my work was terrible, even when it wasn't.

Then I had the brilliant idea that I should try crack.

It's amazing that crack only exists because of an economic glitch in the drug trade. Supposedly there was a huge glut of cocaine in the early 1980s, which made the drug, which used to be expensive and therefore very popular at Studio 54, insanely cheap. So some enterprising dealers decided to convert the powder to a rock form that you can smoke, and sell it in small quantities that regular people can afford. The rock was also purer than the powder, so you got a better high. Wasn't that thoughtful?

By the time I decided to do crack, it was no longer all the rage as it had been a generation earlier—local dealers weren't even bothering to shoot one another anymore—but then I've always been a late bloomer.

After a night of drinking in April 2009, I ended up in someone's apartment in Harlem. Like anyone who was breathing during the 1980s, I'd heard the term "crack house" a thousand times, but I'd never really given it any thought. Even now, I didn't know if I was in one. But there were maybe a dozen people, and every one of them was smoking crack. Does that count?

There were no conversations, no small talk. It wasn't clear if any of these people knew one another, or cared. Every hour or so, a guy would come to the door with more rock.

It didn't take long for me to understand why no one could be bothered with anything but the pipe they were sucking on. Take the greatest orgasm you ever had, multiply that by five, and then prolong it for eight hours, that's how good it is. Maybe better. I could see why people get hooked on it.

I did look up between hits to see a beautiful, large-breasted girl who was missing a front tooth take her top off.

One of the other guys there told me, "She doesn't have any money for drugs, so she goes to parties, takes her top off, and people let her stay and do drugs."

There were a bunch of men in the room, but no one even tried to touch her. That's what crack does, makes you think only about the drug. Nothing else matters, not even a beautiful half-naked woman.

At one point the girl stood up in the middle of the room and started screaming, "I don't have to do this any fucking more! Reason 'cause a that, I have filing skills, and I don't have to do this!"

I said, "Why don't you put your shirt on?"

"Why?"

"I don't know, you don't seem too happy at this moment."

It wasn't that I was a good guy, I just knew what it was like to believe you had no worth.

During a lull while we were waiting for another dealer to show up, this enormous guy asked me to go out on the porch with him. I didn't want to know his name, but he recognized me from TV and decided to watch out for me. He told me his mother was a pastor at a church in Harlem. Then he began this wistful speech about how he wanted to own a $2,000 suit one day. He'd always wanted to go to the opera, but he needed the proper clothes. He vowed that one day he would have a big enough score that he could get the clothes he wanted and go to the opera. "You see," he said, "I have a thing about flutes."

I don't know how many hours I'd been there before I saw a piece of paper taped to the wall. Someone had scribbled in pencil the first line of the Saint Francis prayer, or what they refer to as the Eleventh Step Prayer, the same prayer I had heard the guard recite at the jail in the Bahamas: "Lord, make me an instrument of Thy peace."

God put that there for me. I thought I would never get high again in my life. Then someone handed me the crack pipe. I took a hit.

I probably slobbered on that pipe for another hour, and then I went home. I had the means to do crack for the next fifty years, but there was something about a note with the fucking Eleventh Step Prayer taped to a wall of a crack house that did finally make me leave. *I've lived my entire fucking life in error. Maybe nothing has worked for the last fifty years because I've been living the wrong way.*

As my enormous caretaker put me in a cab that day, he said, "I'm not a crack addict. I'm a crack user." He reached out his hand to shake mine. His grip was firm, reassuring. Then he said, "Can you lend a brother twenty dollars?"

When I got to my apartment on the Upper West Side, the first thing I did was take a shower. I hadn't bathed since the day before, and I'd been doing crack all night. You sweat when you drink and smoke crack. I do, anyway. As it happened, I had an appointment that afternoon with my psychiatrist, so I chain-smoked for a couple of hours and thought about this whole revelation until it was time to see her.

When I got to her office, I told her what I'd done.

"Do you think you'll do crack again?" she asked.

"I don't know." I really didn't. I knew it wasn't a good idea, and the prayer on the wall had really shaken me. But it felt amazing while I was smoking it.

She picked up the phone on her desk and said, "You're going to the hospital tonight." She started making calls to arrange it.

I nodded. I wasn't at all surprised. In fact, I kind of welcomed it.

In a few hours, I was uptown at New York–Presbyterian Hospital on 168th Street, which has one of the top four psychiatric departments in the country. The doctor I saw there wanted to put me in the psychiatric lockdown unit, where you wear special slippers and they drug you. I realized that I wanted that, too. I wanted residential living with a doctor around all the time. I was prepared to live in a medical setting and be given high doses of tranquilizers every day for the next however many years until I died. That's how low I was.

But when they took me to that ward, I got recognized really fast, and several of the patients became agitated.

One guy approached me, raking his hands through his hair like he had fleas, and said, "You're Darrell Hammond, right?"

"Uh, no," I said cleverly.

"Bullshit! Don't fuck with me!" he screamed in my face. He kept raking his left hand across his scalp. "Do an impression for me! I'm in a fucking psych ward! Do Clinton!"

I backed away. I was seriously not in the mood for this kind of shit.

The administrator said, "I don't think this is going to work."

They took me back to a regular floor, but the room they gave me was surprisingly luxurious. I remarked on it to the nurse who was getting me settled.

"Do you know who had this room before you?" she said cheerfully. "President Clinton!" Apparently it had been his room when he had his 2004 quadruple bypass. You can't make this shit up.

I spent a couple of weeks getting it together in the Clinton ward, and then they released me. I had missed only one episode of *SNL*, and I made it to the last show of the season. It would be, after fourteen years, my last as a regular cast member. Finally.

On May 16, 2009, Will Ferrell, who had left the show in 2002 to pursue his film career, came back to host, which was a fantastic opportunity to bring back some of his best recurring characters from the show. Will had recently starred in his one-man Broadway show *You're Welcome America: A Final Night with George Bush,* which had gotten enormous positive press coverage. So of course we did the cold open together, me as Dick Cheney, and Will as W. I don't know if the audience could tell, but I was on so many medications, and so disheartened by what had just happened, that I could hardly remember what Cheney sounded like. We got several applause breaks, but I suspect it was more for the audience's delight to see Will's W back in action. I did get to say "Live from New York" one final time.

The writers brought back *Jeopardy!* for Will to do Alex Trebek. Tom Hanks did a cameo as a brilliantly clueless version of himself during the sketch—he kept speaking into a pen, thinking it was a microphone— and I reprised my Sean Connery for the sketch, which I hadn't done in four years, since Will's previous stint as host. The special surprise was Norm Macdonald resurrecting Burt Reynolds after a decade.

At the good-byes at the end of the night, I went out onstage for the first time since McCain hosted. Out of character, I barely knew what to do with myself. In the early years when I'd gone on stage for the good-byes, I usually stood to one side or in the back. Now Will had me stand next to him out in front. While he thanked the various guests, Norm stood behind me and draped his arms over my shoulders to give me a big embrace. Then Artie Lange gave my shoulders a squeeze. Finally, Will put his arms around me and said, "This is Darrell Hammond, right here," as Bill Hader and Norm pointed to me. The cast was lovely and embracing. Anne Hathaway, who'd had a cameo in the final sketch, "Goodnight Saigon," hugged me. Standing among all those people, I felt each person's hands as they reached out to touch me on the back while the credits rolled. It was as lovely a moment as I'd ever had on the show. Afterwards, I was talking with my friend Julie when Tom Hanks came up and told me it had been an honor to work with me.

It was never formally announced to the public that I was leaving, and the media noted at the beginning of the next season that I had mysteriously disappeared from the opening credits.

For some reason Lorne had me back for six more appearances the following season. Three of those appearances were on the new edition of the show, *Weekend Update Thursday*, and the other three on the regular Saturday show. They all went reasonably well. In fact, after my appearance as California governor Arnold Schwarzenegger trying to sell the California wildfires to raise money for his beleaguered state, Lorne called me and said it was as good as anything I had ever done on the show, and it might well have been.

Strangely, I had risen from my darkness, a darkness where only the past and future live, and settled down in the present. I had been on my game. When it was over, the crowd cheered.

And yet, thirty seconds before my penultimate appearance in the Hall, a few weeks later. . .

In my mind's eye I see the hibiscus bush and hear it thumping against the kitchen window of the house on Wisteria Drive.

Thump thump.

See the pretty people, the important people, the wealthy people, people with pretty skin and flat bellies.

Love your scarf. Is that from Bergdorf's?

No, I don't buy my own clothes. I use a professional buyer. Alicia. Do you know her? Here's her card.

Thump thump.

Overpaid people, androgynous people, Goth people, underweight people. The Hall is electric. We are going to air.

Thump thump.

Is that Anne Hathaway? God, she's beautiful. No, Anne hasn't been here for a while. Certainly looks like her, though.

Thump thump.

Oh my God, is that A-Rod? Oh my God, it is A-Rod.

Thump thump.

I hope this one goes well. Where is Lorne? Have you seen Lorne?

Thump thump.

He's with Tom Hanks, but he'll be here, you know that. Five seconds, Darrell.

Thump thump.

Lorne walks into the Hall. Why can't I talk? My vocal cords hurt so bad.

Thump thump.

Marci Klein walks into the Hall. I'm running out of breath. I'm not my age anymore. I'm no longer in the Hall. I'm back in the house on Wisteria Drive.

Thump thump.

Stand right there. You love Mommy, don't you?

Yes, I love you, Mommy.

Then hold your hand here, on the doorsill. Right there. Like that.

Thump thump.

A-Rod, my hero, twenty feet in front of me. Kate Hudson on his arm. The room is flashing red. I grab my wrist to steady myself. Did my thumb leave an indentation?

Thump thump.

A-Rod with a quizzical expression. He is looking at something he does not understand: malfunction. A-Rod does not understand malfunction because A-Rod was not designed to malfunction. A-Rod was designed to hit eight hundred home runs and do big things.

Thump thump.

Gena: Five. Four. Three. Two—

The Hall is red. Marci is beautiful.

Thump thump.

Seth Meyers: Here to comment is Lou Dobbs.

Thump thump.

My thumb has left a mark.

It's two years since my last show. I don't get recognized as much as I used to, but sometimes I do. I went into a Sephora store recently for moisturizer. (Shall we pause for a moment while you finish chuckling about that?) Of course everyone in the store was disturbingly attractive.

Oh, shit. All of you make me feel really bad.

One of the young women came up to me.

"Excuse me, sir."

Shit. That word.

"Are you Darrell Hammond?" Her bright-eyed, eager smile told me she was more likely to offer to help me cross the street than give me her phone number.

I nodded. As flattering as it is, I find this part of my career a little embarrassing.

I'm in my bedroom closet in Melbourne, drinking myself into a stupor at fifteen; the psych ward at New York–Presbyterian; a Harlem crack den.

"Ohmygod, I've loved you for fifteen years, since I was ten!"

I know everyone who works the night shift in the emergency room at New York Hospital on a first-name basis. Have you seen my arms?

These lovely people laughed at me on national television every week—that's all the fans knew. A generation grew up watching me. Impossible.

CHAPTER FIFTEEN

The Golden Years Redux

Melbourne, Florida
2007

About a month before my father passed away and before he went into hospice, I thought it might give him a thrill at the end of his life to see his son hit some baseballs. I was fifty-two years old and hadn't swung a bat in thirty years.

I got in touch with my old high school friend Wayne Tyson—who had taken my spot on the junior college team when I got sick—and asked him if he'd pitch to me. He agreed, and offered up the field at Palm Bay High School, where Wayne was the coach of the Palm Bay Pirates baseball team. Once again, I flew down with Eddie.

On the day, Wayne went out to the mound and pitched like he'd pitched ten thousand times to me when we were kids. Out on a field, that soft breeze coming off the ocean, the gardenias and jasmine in bloom, the smell of the grass and the dirt, I was the fucking Mick again.

My father sat on the bench in the dugout. Wearing his baseball cap, leaning on his cane, he watched the proceedings as carefully as he ever had. The night before, we had watched the Pirates in action, and my father commentated the entire game. He could barely get around anymore, but he hadn't lost his encyclopedic knowledge—or his love—of baseball.

I took a few practice swings before I stepped up to the plate to get the feel of the bat in my hand. I wiffed the first couple of pitches, and then it all seemed to click, long-dormant muscle memory waking up.

For a few minutes it was transcendental as the bat made contact with pitch after pitch. I might as well have been wearing my old Howard's Meats uniform and had long yellow hair. I was in some kind of a dream. I started hitting these shots I had only fantasized about for decades. And then I lifted one down the left field line, a high, high fly ball—it seemed to stay up in the air forever. It fell a few yards short of the wall.

Out of the corner of my eye, I saw my father struggle to his feet. Slowly, he hobbled onto the field, leaning heavily on his cane. He shuffled over to where the ball had landed and paced it off, like he used to do when I was a kid. The fence down the line was 320, a fair bit farther than the 297 at Wells Park.

After Eddie, Wayne, and I gathered up the bats and balls, and we were walking back to the car, my father paused. Looking me directly in the eye, and, for the first and only time in my life, he held my gaze for a long moment and said, "It woulda gone out."

THE LAST CHAPTER
I Mean It This Time.

The Sanctuary
Westchester County, New York
2010

I have to admit that dredging all this shit up to write
this book pretty much kicked my ass. After the episode in the crack house and my last hospitalization, I
thought I'd made peace with everything. It turns out,
I hadn't.

Thanks to my dear friend D, I'm still alive. She found
me when I was in the depths of a relapse with a beer in
one hand and a large carving knife in the other. The
night before, well into this drunk, I'd made the profound mistake of going to see my daughter, and the look
on her face when she saw me in that state shamed me.

After wrestling the knife away from me, D called the
ambulance and rode with me to the emergency room

at New York Hospital. I *knew* I should have gotten an E-ZPass for that place.

I was in the psych ward overnight. While I was there, a male patient walked back and forth in the hallway, a cell phone to his ear, and he kept saying, "I'm sorry, Mama, I think bad things."

The doctors were sufficiently concerned about me to send me to the Sanctuary, which is one part psychiatric hospital and one part rehab. Over the course of my stay, I spent time in both.

We had quite a little society at the Sanctuary. There are fourteen wards, divided up mostly by ailment. I was on a ward reserved for boldfaced names: a couple of world-class athletes with ruined careers; an international beauty pageant winner; European royalty; and some Middle Eastern goddess of considerable power, although I was never able to determine what it was. A world-renowned actress had checked out just before I arrived.

There was a ward dedicated to folks who were in rehab by order of the court, as a condition of a sentence or parole. That's where the most interesting person in the whole hospital was. The Mayor, as we called her, was a captivating young thief with full lips, a ring in her nose, and she always wore something that displayed her rather splendid midriff (she wouldn't have gotten away with that on our ward, but on the criminal ward where she was, they didn't care). Think a young Molly Ringwald. She had just had fifty-two stitches removed from her mouth from injuries sustained fighting at the women's prison at Rikers Island in New York City, where she'd been incarcerated after she was arrested for felony narcotics and armed robbery. I adored her.

To be clear, there was nothing sexual about our relationship, we were just drawn to each other and spent every minute we could together. She'd already been there for a month or two when I arrived, and she seemed to know everybody, both patients and staff. She had this astonishing ability to figure out what was going on with people, what it was that mattered to them, in a very short amount of time, and then feed that every time she saw them, and tie it up with a ribbon in the form of a funny line when the conversation ended. She walked up and down the hall, utterly charming everyone. She could have been the head of the PTA. You might find her chatting just as comfortably in Spanish with the custodians as in English with the likes of me. She'd leave little notes for people. One minute she was asking after one patient's granddaughter, the next she was talking to a security guard about how his family back in Haiti was doing after the earthquake. She knew all the rules, and how to get around them. Will there be a strip search tonight? Ask the Mayor. Are they testing urine today? Ask the Mayor.

One of the patients was a soldier who'd been shot in the head. He had lingering mental and physical impairments, and he was struggling to make some sense of what had happened to him, if it had been worth it to be shot in the head. He couldn't stop nodding his head up and down from the neurological damage he'd sustained, and he was depressed.

Sensing this, one day she approached him and said, "What is your favorite band?"

He looked up warily from the crack in the floorboard he'd been studying for the last half hour. "Lynyrd Skynyrd," he said.

She said, "There wouldn't be a Lynyrd Skynyrd if someone hadn't been willing to do what you did."

He brightened visibly. For the rest of the time I was there, he was a slightly different person.

There was a pretty black girl who'd been raped repeatedly. She wasn't able to hold conversations. When you said hello to her, she answered, "I'm Delonda Jackson," and then she'd walk away. She spent a lot of her time wandering up and down the hall, stopping to clean her hands at the Purell dispensers, talking to herself. And yet there was a sense of calm about her, like a soldier standing before the firing squad who has accepted his imminent execution. There was also a sense of fierce pride in her too; she would face that firing squad with her head held high.

One day the Mayor had a friend bring her a boom box and some CDs. Somehow the Mayor persuaded Delonda Jackson to dance with her. Days later, Delonda Jackson was still moving to her groove. It was wonderful to see her experience such joy. Late one evening I wandered down the hall in search of sugar for my coffee, and I came upon Delonda Jackson and the Mayor dancing with that veteran, in his pajamas and bare feet and the company hat he always wore.

"Honey, I know what you're thinking," she said one day when she found me pacing the hall with a furrowed brow. "You feel like there's a universal balance sheet, and it ain't balancing out in your favor because of things that happened to you and the life you've had to lead. And you want to be paid back. But you ain't getting paid back, and you need to stop thinking that you will."

It was a huge breakthrough for me.

If you're an alcoholic, you're not like other people. You have a progressive fatal illness. Trauma is also progressive and fatal. So I had two of those, and for most of my life they debilitated me. More than a few times, they nearly killed me.

The first thing they told me at the Sanctuary was a phrase from AA's Big Book: To live life on a spiritual basis or die an alcoholic death are not always easy alternatives for an addict to face. That was a thunderclap. Addiction had turned me into someone who would rather be dead than be forgiving. A disease as lethal as Ebola, it kills its host in messy, horrible ways. But the love of my daughter and the time that we spent together was making me into someone who felt real human emotions. I'd blown it, though, and I wasn't sure she'd ever speak to me again.

Prior to my going to treatment this time, you could not on any level, with any device, apparatus, or mechanism, have convinced me that I was playing a part in my own self-destruction. I felt entitled to be angry, I felt entitled to be self-destructive. I became a monstrous figure, someone who was willing to scare others to get them to lay off. If I got drunk, I'd be willing to cut myself.

So here I was atop this volcano of impotent rage, and getting sicker and sicker and sicker, and feeling like a victim. But feeling like a victim tends to make a person selfish, to play Pin the Anger on the Donkey. A relatively innocent bystander can become the object of your scorn. I demanded that you play by my rules, and I also demanded that you guess what they were. Therefore, just like the Mayor, I never had real relationships with anybody. It was easy for someone who was beginning to care about me to become upset with me. Once they became angry enough, I could say that the world was a hostile place, and then I would drink. It's like my brain said, *The only thing that works is this alcohol shit. C'mon, let's get it.*

The doctors explained to me that my disease was one of mistaken beliefs. We have to believe things that are enormously untrue to continue with the addiction, chiefly that our behavior doesn't really have any effect on others. And if it does, so what? I wasn't willing to acknowledge that I could be an asshole or that anyone was suffering from this but me.

Until I saw fear in my daughter's eyes.

What you find in rehabs, even a really good one like this, is that the counselor with the least ability is the tough-love counselor. That's how they position themselves. They call it carefrontation, but what they actually do is say mean shit to you. Not perceptive or analytic, just mean. The trick was to get the patients to acknowledge ugly truths about themselves so that they can heal, but do it in a way that doesn't traumatize them. The tough-love counselor didn't have that skill. Everyone else on staff was really gifted, but not her.

One night I called her a bully.

She said, "You're rude. You want your daughter to be like you?"

"If my daughter thought she was being picked on by a bully, I would want her to be like this, yeah."

She was also in charge of who got urinalysis and who got cavity-searched. She was always making the Mayor pee in a cup in front of her so she could see the pee exiting her body.

One night in Group, in front of everyone, the tough-love counselor on our floor carefronted, aka humiliated, Juanita, a voluptuous twenty-five-year-old girl who had been living with a sixty-year-old man who supplied her with crack in exchange for her favors. She was not allowed to leave the house or have friends. Juanita wanted to be a graphic artist, to go to a sober house, to get her five-year-old daughter back. It was a miracle she'd gotten herself into rehab. You have to be gentle with someone who's been traumatized, and the tough-love counselor got it all wrong.

Juanita flipped out and walked out of the room. We could hear her at the pay phone calling the sixty-year-old, screaming and crying. Within a few minutes, she had packed her meager possessions in a duffel bag and left. We looked outside and saw her get in the car. It wasn't difficult to figure out what was going on in there.

One patient, who'd recently been paroled after a twenty-year stretch in prison, said, "Damn! Mother-fucker is gonna get some twenty-five-year-old pussy this eed'nin'!"

Everybody but the Mayor started laughing. Evidently the sixty-year-old had brought drugs, and she was making her first payment for them right there in the front seat.

I heard that the Mayor walked up to the counselor and said something like, "You're a disgrace to people who are trying to change their lives. You can't be abusive to someone who's been through what that girl has been through. If you were any good, you'd know that."

Later, I said to her, "Wasn't that a little harsh?"

"That girl is gone. She could OD on crack any time. This woman has injured a patient, and that patient is fucking gone."

All of the addicts there were looking for a reason to bail. One of the doctors compared us to people on a relay team waiting for the baton to be given to us. We welcomed anything that could push us into victimhood so we could hate and drink and use.

There was one twenty-something kid named Arthur who'd been traumatized so badly he couldn't speak at first. When he did begin to talk, the Mayor gave him an audience all the time. He was friends with Juanita, and when she got in the car with her captor, he was heartbroken. The Mayor spent a lot of time with him that night, explaining that he would find another friend one day.

There was one patient the Mayor couldn't get to, an eighteen-year-old Angelina Jolie lookalike with cascades of shiny hair. She wandered around in tiny tank tops and low-slung pajama bottoms with no underwear. I saw her walk up to one patient on methadone maintenance and say, "Come party with me." She took him outside into the dark night to do whatever he wanted in exchange for some of his methadone. At first I thought people were taking advantage of this young thing, until I realized it was the other way around. She would let the women feel her up in order to get stuff from them too. Some people were on Ativan, some people were on Klonopin, some were on Seroquel. Some of the patients cheeked their meds, pretending to take them, then spitting them out when the nurses weren't looking to use as currency later.

She walked into Group one day fully stoned and said, "I'll suck any dick, I'll eat any ass, I'll even fuck that fat piece of shit." She pointed to a minister who did volunteer work for addicts and criminals.

One night she told me she wanted to give me a Steaming Carl.

"First of all, you're eighteen, and my daughter's almost a teenager," I said. "And second of all, I'm almost embarrassed to ask this, what is a Steaming Carl?"

She described an act that would have been at home in a German *Scheisse* movie.

"Okay, that would be a no."

Why is it that all the seeming goddesses who have hit on me throughout my life are batshit crazy?

The staff thought it would be nice to bring the patients from all fourteen wards together for a karaoke night. Everyone was herded into an enormous hall. There were about twenty tables, each covered with bowls of "fun size" candy bars and bags of chips—every form of sugar or artery-hardening substance known to man. They brought in a technical team to set up a big video screen with the words to songs written across the bottom. You haven't really lived until you've had karaoke night in a psychiatric hospital.

As people mingled with one another, they would introduce themselves by saying their name and which ward they were from: schizophrenic, borderline, manic-depressive, depressive disorder, eating disorder. There were some really good singers, including one bulimic woman who was all bones but for her beach-ball belly; she was pregnant.

Then there was Tony, a Johnny Depp lookalike with an enormous phallus (wait, there's a point to this) and brain damage from years of meth abuse. He was the sweetest guy, always doing little favors for people. "You need milk in that coffee, man? I could get you some milk. I could get that." He'd also expose himself to the other patients in exchange for their meds, which everyone seemed to find hilarious.

Tony occasionally saw people who weren't there. He'd walk across the room to say hello, then have to pretend that he'd gone over there for some other reason. He'd grab his dick, then stroll back to his chair like John Travolta at the beginning of *Saturday Night Fever*. Tony tried to sing "Respect." When he got to the part, "R-E-S-P-E-C-T," he became flustered. He walked away from the microphone to say hello to an imaginary visitor, grabbed his dick, then came back to the mic after that part had finished and resumed singing.

But honestly, it was a wonderful night until someone put Michael Jackson's "Billie Jean" on. All at once, it dawned on the patients that he had died not long before of a horrible overdose. Near-death experiences were what connected us all. People at the different tables, with their different maladies, began to get upset in textbook ways, each according to his own affliction.

We had all been triggered, and it quickly turned into a melee of screaming and crying and twitching.

The nurses came rushing in like a mom saying, "Everybody out of the pool!" to herd the patients back to their wards. They spewed the little catchphrases they were always saying to the patients.

"Easy does it."

"Fake it till you make it."

"It's not about getting out of the rain, it's about learning how to dance in the rain."

I never did find out which one of the lunatics chose that song.

One night toward the end of my stay at the Sanctuary, the eighteen-year-old in the pajamas and a thirty-year-old stripper she palled around with decided we should play Would You Rather . . . . They wrote out suggestions on scraps of paper, and we were all given a few to read aloud. Each one was more foul than the last: assholes, cocks, balls, vaginas, camel toes, blow jobs, golden showers, fisting, diarrhea, cum, dildos, bestiality, erectile dysfunction, gangbangs, bologna, incest, cock-chugging, it was all on the table. (I have to confess I introduced "cock-chugging," which I'd gotten from Eddie. It was an immediate hit in the criminal ward.)

In the middle of it, Arthur the trauma patient blurted out, "Would you rather be raped or be a rapist?" He stood there looking at the ground, his hands shaking violently, his mop of brown hair covering his face. The Mayor talked quietly to him until he stopped shaking.

Tony kept going up to the Mayor and saying, "When am I gonna hit that shit?"

"You don't want to hit this, honey," she said. "It's too much for you. This is for the big boys."

Not long before the Mayor was due to check out, she sat me down to talk. I knew what was coming.

"We can't be friends on the outside," she said. "I'm a criminal. I like to steal, and I'm a junkie."

I asked her why she liked to steal so much.

"I never had a chance. I was kidnapped. My dad was a drug addict, he OD'd. I was left on my own in Mexico. I never had a chance to go to school and join clubs and date. I didn't have a childhood, and it wasn't my fault. I'm owed. I go to Wal-Mart to practice, but Wal-Mart is dick. I like to steal from the big people. I fucking deserve it."

What allegedly got her into the Sanctuary was that she had stolen countersurveillance equipment from the NYPD, and the CIA had arrested her. She had no intention of getting better. She'd only come here to avoid prison.

My doctor told me, "You know she's a sociopath, don't you? You can't catch her in her cons. She feels these things when she says them, but she doesn't intend to have a meaningful relationship with anybody."

"Did the CIA really arrest you?" I asked her.

"Yes."

A rather sobering answer. If she really had been arrested by the CIA, that seemed like a deal breaker to me. I will hang around just about anybody, but that was too much. And if it didn't happen, and she'd made that shit up? Hell, I think that would be even worse, wouldn't it?

A few days after the Mayor was released, she called me on the cell phone I'd eventually been given permission to have. She had driven her car into a tree, she was high, and she'd stolen someone's stash, which was apparently in the car with her.

While we were on the phone, I heard her say, "Officer, what am I being booked for?"

"Leaving the scene of an accident, felony narcotics, and DUI."

I knew she was on parole with the condition that her urine tested clean. She was fucked. "I want you to forget you ever met me. I'm going away for a long time."

I have to admit, I was a little bit crushed, but I wasn't surprised.

A couple of days letter, a package arrived from her with a pair of Nike sneakers and a note that was signed with a number 7. I guess I must have told her how obsessed I was with Mickey Mantle.

Dr. K. looked at the note and said, "She's good. But I'm not going to let you keep the shoes."

"Why can't I keep the shoes?"

He thought she was capable of planting a chip in the shoes. "She's going to find out where you live, and one day she's going to come for you." I didn't keep the shoes.

Dr. K., the Sanctuary's resident genius, a small Moroccan man with a disarming smile who the world's power structure sent their fucked-up children to, explained to me that when you have been tortured and beaten by your mother, when you've had suicides in your life by people you'd exchanged I-love-you's with, and you've been cursed with a progressive fatal illness you didn't ask for, your brain starts searching for a way to explain it. Most of the time, your brain says, "It's because of you. That's why your mother hit you, cut you, slammed your hands in the door." You think you're shit, you think you're worthless, you think you're unlovable, you think you can't do this, you can't do that. Life is always bad. Your brain has tried to simplify a perfect storm because it's so confusing. Why would this shit happen to me?

One day early in my stay he came to my room and said, "Do you believe you are the kind of person who is willing to frighten the people who care about you for the sake of your anger? Are you willing to sacrifice your life for it? Are you willing to take this all the way to the grave?"

As he stood in my doorway to leave, he said, "I would suggest to you, Mr. Hammond, that you are."

What must it be like to wake up in the middle of the night unable to find your husband, knowing that he's been in the emergency room with cut wrists before? It must be terrifying, especially when there's a child. I don't know that you ever fully repair trust. How do I make amends for the way I behaved when I believed that I was entitled to everything because I was mistreated?

After the doctor left my room, I looked out the window into the bleak snowscape outside, no friends, no family, no money. My credit cards and my phone had been taken away. I was completely cut off from my life. A few years earlier, Drew Barrymore had treated me to a burping exhibition. Sean Connery told Jay Leno on *The Tonight Show* that I did him better than he did himself. Gwyneth Paltrow held my baby daughter, and Kate Winslet taught her how to pucker up for a kiss.

How in the same lifetime could I have had Tom Hanks telling me he'd been honored to work with me on my final show? Compliments from Bono and John Mayer. Lengthy discussions with David Duchovny and Sylvester Stallone. Tom Brokaw sharing a story with me in the hall about visiting the beaches of Normandy. Meeting John Kerry and mistakenly introducing him to someone as Trent Lott. Hanging out with a shirtless, cigar-smoking Chris Matthews in his backyard while he talked about his favorite photo of Babe Ruth. E-mails from Maria Bartiromo, Katie Couric's voice on my answering machine, asking me to imitate Matt Lauer as a favor. Hell, I'd kissed both Paris Hilton and Monica Lewinsky in sketches.

These things kept whizzing through my mind as I sat there, filled with the first healthy shame of my life, with no way to contact anyone, no wallet, no e-mail, no connection at all. How could I have lived in that world and lived in this one as well?

In *Les Misérables*, Jean Valjean is imprisoned by a sadistic guard for nineteen years for stealing a loaf of bread to feed his sister's starving child. He's tortured, beaten, put on the rack, whipped. When he comes out, he determines that he is going to take his rage out on the world. He meets a priest one evening who invites him in to stay for the night and get out of the cold. The priest gives him wine, bread, and meat, a nice place to sleep. Jean Valjean repays the priest by getting up in the middle of the night and stealing all the priest's silver.

The police catch him and bring him back to the priest. They say, This criminal has your silver. And the priest says, I gave it to him. I gave him this gold, too, which, he says to Jean Valjean, you forgot. They leave, and the priest says to Jean Valjean, Remember this, my brother, the deity has a higher plan. Use this precious silver to become an honest man. By the witness of the martyr, by the passion and the blood, God has raised you out of darkness. I have bought your soul for God.

Jean Valjean goes off into the wilderness. A little orphan girl comes into his life. The process of loving a helpless little girl, trying to protect her and raise her, changes him. He becomes successful, a good father, and years later he runs into his oppressor, Javert. Jean Valjean has the opportunity to kill Javert, but spares him.

Javert is the old me. Rather than feel any sense of forgiveness, accept any act of kindness, or for one second think that he might have been wrong, he'd rather be dead. And he kills himself.

My brilliant psychiatrist Dr. K. had told me that the only way to move on from trauma is forgiveness, giving up the right to hit back harder than you were hit in the first place. To get well, I had to drop the indictment against the people who had hurt me. For my daughter's health, I had to drop the indictment. For someone who has spent a lifetime conjuring up revenge fantasies, being told that I had to walk away without any kind of payback was like watching an endless loop of *NYPD Blue* in which the bad guy always gets away with it. These people wronged me, but I was going to have to change my molecules, change my brain, in order to get well. That's a tall fucking order. It took years and years of hospitalizations, treatment, therapy, and psycho-pharmaceuticals, but I think I finally got it.

When the doctors at the Sanctuary were done with me, they sent me to a dedicated addiction treatment institution in the middle of nowhere, Pennsylvania, for another eight weeks of inpatient treatment. It was the longest winter of my life.

At this second rehab, they said, "Can't we show you ways that you have been paid back? Can't we show you ways that the universe has since been good to you?"

They would have us sit down and make out lists of our blessings and all that kind of shit.

What had a more profound impact on me, though, was all the talk I heard about the place being haunted, that there was a ghost that came out at four-thirty in the morning. The place had previously been a hospice, so a lot of people had died there. "I've never seen it," one security guard said. "I've heard it. I've heard all kinds of things I can't explain."

After growing up in a house my parents had convinced my sister and me was haunted, I couldn't pass up the opportunity, so I got up one morning to see this ghost. Of course there was nothing there.

The next night I woke from one of those dreams that seems so real that it takes a minute or two for you to discern that it wasn't. I dreamed that I woke up and there was a little girl, not more than three or four years old, standing outside my window in the snow. It was dark, and she was crying and afraid. There was something familiar about her, about her eyes. And then I realized they were my mother's eyes. It was my mother as a little girl standing in the snow, shivering and helpless, a pure human being before someone did to her what she had done to me. I had this inescapable sensation that someone had hurt her and continued to hurt her for a long time. More importantly, she had once been innocent.

I realized then that the difference between my mother and me was that I had had Myrtise, who for a few precious years held me and loved me, and my mother had had no one.

When I woke up, the world seemed in focus for the first time in my life. After years of haziness, I had crystal clarity. I touched the wall to see if the plaster was real, the bedcovers to see if they were really made of cloth. And I felt that shame again. I realized my life was shit, and I had participated in making it so. The horror of being without my daughter had been partly my fault. I was capable of being an asshole, and had been an asshole, a self-destructive, mean-spirited prick. I had become the driver of the bus that had been hijacked in my childhood by my parents and escorted me to hell. I understood that I was wrong.

And then I felt like something approaching being whole again. I was ready to go home.

A few days later, after three months of inpatient treatment, I was finally released. Driving away, the grounds a canvas of white from that winter's record-breaking snow, I looked back one more time to see if the little girl was still standing by the window. She wasn't.

I was free.

THE REAL LAST CHAPTER

Honest

New York City
May 2011

"Just when I thought I was out, they pull me back in."
—MICHAEL CORLEONE, *The Godfather III*

Okay, what I just said about being free? That's bullshit. Nobody's ever really free of their baggage. I mean, I'm not in jail and there are no more flashbacks, no more nightmares, no more cold sweats, no more cutting, no more heavy pharmaceuticals. I smile from time to time. I still wear a lot of black, but I'm thinking about going clothing shopping soon.

And these days when I go on an audition, I really enjoy the process—television series, films, even Broadway. Whether the parts or the shows pan out or not, I'm having fun out there. Why is this happening now?

This might be it: I got a new psychiatrist, Dr. Ramsey, who said, "Do you hear voices?"

"No."

"Do you see things? Do you think people are there, looking around the corner at you?"

"No."

"You're not psychotic. You're not schizophrenic. And you're not manic-depressive. You don't need the medication."

So I'm not on those soul killers anymore. I live by the law of mutuality that is found in twelve-step meetings. What we can't do alone, we can do together, that sort of thing. There's something about being in a group of people just like me, trying to recover, hoping to improve their spirits, reaching out to one another, relating to one another. I know that after a meeting, I'll go home feeling good. I once called the office of the head of the New York Psychiatric Institute when I was still drinking. I told him I was in a twelve-step program, and I wanted a referral for a psychiatrist. He said, "Do you know that's the best cognitive therapy ever invented?"

What else? I go to the gym now. Every Wednesday is Guys Who Used to Have It Going On Day. We sit around on our towels, taking a steam, thinking, What the fuck happened, man? We're still relatively well built, you can tell that once we might have looked pretty good, and we did pretty well in life; we've achieved a lot of what we wanted to achieve, seen a lot and experienced a lot, and now what do we fucking do? I wish I could go to every high school and give a speech, "Just in case you're interested, it's never going to be better than this."

Although lately I've been feeling that might not be true. I hasten to emphasize, *might* not be true. I'm willing to consider the possibility that things do get better.

It had been more than a year since my last guest appearance on *SNL,* and I'd grown comfortable with the idea that I had finally moved past that chapter of my life and on to new things. I was gearing up to play Truman Capote in *Tru,* a one-man play by Jay Presson Allen scheduled to run for the month of June at the Bay Street Theatre in Sag Harbor, New York. I had been part of an ensemble cast in David Mamet's comedy *Romance* the previous summer at Bay Street, but *Tru* was a whole different ball game: I was the one and only cast member, and it was a serious drama based on Capote's own words—ninety minutes of them. Plus I had big shoes to fill in playing the man known as the Tiny Terror: Robert Morse had won a Tony Award for Best Actor for his turn in the original Broadway production in 1990 *and* an Emmy for the television version that ran on PBS a year later.

For me, the real challenge was learning how to be Truman Capote, who was eight inches shorter than me, and had a voice so bizarre and high-pitched that it had sounded fake in real life. If you listen to a recording of him with your eyes closed, you'd think you were listening to Granny from the old Tweety Bird cartoons having a stroke. Figuring out how to play him without making him sound like a caricature—as I'd done with almost every character I'd played for the past sixteen years—was as daunting as learning how to hit one of C. P. Yarborough's curveballs. Fortunately, I had had the luxury of eight months to study and learn him—a far more comfortable proposition than the four to six hours I had to learn most of my *SNL* roles. And coincidentally, or maybe not, my addiction history was going to be an asset for this role: over the course of two acts, a despondent and lonely Capote becomes increasingly wasted on martinis, Valium, pot, and cocaine. It was entirely fitting that I had conducted part of my Capote research during the previous winter's rehab extravaganza.

For much of the spring, I spent my days on the second floor of the New 42nd Street Studios building in Times Square, an elaborate new performing arts rehearsal space. With the New York City Ballet in a studio on one side, and the Shakespeare in the Park crew working on their summer production of *Measure for Measure* on the other, I worked on blocking *Tru* with my directors, the supremely talented Judith Ivey and Matt McGrath. Working with them was like getting a Yale Drama School education in fast-forward—when Judith told me to stop focusing on Capote's voice and pay attention to acting instead, I knew I was operating in a different realm than what I'd been accustomed to. But it was going well. Once again, my life had taken a turn into the land of the impossible—I was going to play Capote in a major American theater.

And then my phone started to vibrate.

On a Tuesday, three weeks shy of opening night, I received a text from Steve Higgins asking me if I could do Trump that Saturday on *SNL*.

Tina Fey was returning to host, and they wanted to do a sketch about an imagined Republican debate that included Tina as Sarah Palin and me as Donald Trump, who was publicly toying with a presidential run and making a lot of headlines raising questions about President Obama's birth certificate. As a result, Trump had also been the butt of jokes at the recent White House Correspondents' Dinner. CSPAN's cameras zoomed in on his grim face as he endured a thorough roasting by the president of the United States. His scowl became even more severe when *SNL* head writer Seth Meyers, who was the entertainment for the evening, started in. Those pictures would be all over the news media in the days that followed.

In a conversation with Kenny Aymong, the supervising producer, later in the week, I said that I was in the middle of rehearsals for *Tru*, but he said they would accommodate my schedule.

A few minutes later, Lorne's assistant Lindsay Shookus called to give me the details. She agreed to messenger over the initial script.

Saturday morning, I went to the studio, where Judith and Matt put me through my paces until about 2:30 p.m. Then I grabbed a cab uptown to 30 Rock in time for the afternoon run-through. Before I even got in the door of the building, I ran into Mary Ellen Matthews, the genius *SNL* photographer who is responsible for the bumper photos, as the shots of the hosts you see after commercial breaks are called.

"Are you on tonight?" she asked, giving me a warm hug.

"Hey, Darrell!" a voice yelled from behind me. I turned to find Jodi Mancuso, the Emmy Award–winning *SNL* hairstylist.

A moment later, a tap on the shoulder: another Emmy Award-winner, Katreese Barnes, assistant music director and keyboardist with the *SNL* Band.

The warm welcome went a long way toward calming my nerves at being back. I assumed everyone knew what bad shape I'd been in when I'd left in 2009, but it didn't seem that way at all.

I went inside, past the line of tourists waiting in line for the NBC Studio tour, and headed for the elevator and up to the eighth floor.

Upstairs, Lindsay greeted me and showed me to my guest dressing room, where I picked up a new version of the script before heading over to the hair and makeup department, down the hall near the cast dressing rooms. On the way, I passed the shelf where the head models for the cast and recurring hosts are kept. I was a little sad to see mine was gone.

But when I turned into the hair and makeup room, there it was, and on top of it was a masterpiece of blond swirls that rivaled the real Trump's signature "do." Bettie Rogers, the head of the hairstyling department, another Emmy winner, had outdone herself. There was a set of handmade bushy eyebrows, courtesy of Emmy-winner Louie Zakarian, the head of the makeup department, to go with it. After a fairly elaborate process to cover up my own hair, which I'd let grow long for the Hound Dogs pilot, Jodi placed the wig on my head, glued the eyebrows on, and I was transformed.

A few minutes later, I was summoned to the main stage by the voice of Gena Rositano, the stage manager, over the PA system. There, I joined Tina as Palin, Kristen Wiig as Michele "Crazy Eyes" Bachmann, Bobby Moynihan as Newt Gingrich, Jason Sudeikis as Mitt Romney. Each of us stood behind a podium as the candidates would for a debate, although mine was fashioned out of gaudy gold columns. "Unofficially" joining our fake debate was Kenan Thompson as Jimmy McMillan, the eccentric former New York gubernatorial candidate known for his dramatic facial hair and Rent Is Too Damn High Party affiliation. Bill Hader was moderator Shepard Smith.

As we ran through the sketch, a team of writers, along with Seth, who wrote the script, made notes about what worked and what needed to be changed. Kristen Wiig wanted to know if the way she bugged out her eyes when her Michele Bachmann challenged America to a staring contest was okay. Steve Higgins suggested I turn toward Tina when I got to the part of the sketch where Trump talks about how everyone but him is a loser except for Sarah Palin. Steve thought the two-shot of us would be the image that would run alongside the reviews in the Monday papers.

When we were done, I headed over to the costume shop to get fitted for my Trump suit. Nearly every superstar on the planet has been naked or half-naked in this place. The shop is a warren of rooms filled to bursting with clothes. Every wall is lined with racks, and the center is piled with clear plastic bins crammed with bras, shapers, corsets, Spanx, and—my favorite—Man Spanx. I noticed a large padded, pastel-colored bikini-like garment hanging on a wall and wondered what it was for. (Tina would wear it in a mermaid sketch later that night.)

And then it was back to my dressing room, where I had to turn myself into a larger-than-life six-foot-three businessman and reality TV personality with a booming voice from the fey five-foot-two Truman Capote I'd been inhabiting for the previous few weeks. As in the old days when I was on the show, I asked research to find recent clips of Trump speaking so I could see if there was anything new I should be adding (I'd done the impression more than a dozen times before). Over the next few hours, I watched over and over again videos of Trump giving a controversial speech in Las Vegas, a press conference taking credit for making President Obama release his long-form birth certificate, and instructions for a new *Celebrity Apprentice* challenge to Meatloaf and Star Jones.

Around six o'clock, I was really starting to feel the strain of the long day, so I decided to catch a nap on the couch in my dressing room. Whatever else my mother had done, she had at least trained me never to put my shoes on the furniture. As I reached down to untie the laces of my sneakers, I was suddenly taken back to that horrendous day the previous November when I'd taken a large carving knife to my arm. That morning, knowing that I had to come to *SNL* after rehearsing the play, I'd been even more discombobulated than I usually am in the morning. For the life of me, I couldn't find my shoes. Annoyed with myself but very short on time, I reached into the back of my closet and grabbed the first pair of shoes I could find. It wasn't until I went to take them off that I realized that I'd grabbed the same pair I'd worn that cold autumn day: they were still speckled with my blood, now dried to a sad, dark brown.

Around seven-thirty, there was a knock on the door. The young man from wardrobe who was assigned to me brought my suit. This being *SNL*, where every guest is taken care of in every way, he even helped me put the damn thing on.

For the dress rehearsal and the live show, costume, makeup, and hair moves to a space along the main hall leading to the studio to accommodate the fast changes everybody needs to make between sketches. On one side of the hall are the quick-change booths, one for each cast member, and each manned by someone from wardrobe. On the other side, the hair and makeup team assemble all the bits and pieces of their part of the process, moving each cast member through in an efficient if slightly harried assembly line, as if we were bombers coming off the line at Willow Run during World War II. As Jodi put my wig and eyebrows back on, I read the updated script.

After that, I headed over to the area under the audience bleachers where the cue card guys were furiously writing out the new dialogue. I'd taken that walk a thousand times before. Whenever I could, I liked to practice with the cards before going on. Sometimes I'd ask the guys to underline a phrase or sentence that I'd need to emphasize. I practiced facial expressions in a mirror that hung nearby while I fine-tuned the voice.

Then it was out to the floor for the dress rehearsal, the two-hour version of the show done in front of a live audience. Already some of the sketches had been dropped and moved around since the run-through, so the debate was earlier in the lineup and there was less time to get ready. We tried it out with the revised dialogue, and I did the turn to Tina.

Then it was back to hair and makeup to have my wig removed again, and off to my dressing room. They take very good care of the clothes, so my costumer came to once again retrieve the suit. I went back to watching videos of Trump doing his thing.

The dress rehearsal finished at 10:00 p.m. I waited a few minutes before going upstairs to the ninth floor to await the final meeting. There's an anteroom where the interns feverishly compile the changes into scripts for everybody who needs them. While the cast cooled their heels, Lorne, the producers, and the writers were in the conference room talking about what to cut.

Around 10:30, Higgins came out and said, "Meeting!"

That was my cue to pile into the room with everybody to find out the fate of the sketch and talk about whatever additional changes would be made. Steve and Seth and I talked about how to tweak the Trump bit even further.

Steve said, "When you turn to Tina, hold the pose. Don't say anything for a beat before you continue. That's going to be the money shot for the media."

When the meeting was done, the two-hour show had been whittled down to ninety minutes. The cast headed back down the stairs to the eighth floor to get back into wigs and wardrobe and makeup while a posse of young women took over the photocopy room to get the new scripts ready.

One more time, I got the Trump swirl wig and eyebrows and expensive suit on. The GOP debate would be the first sketch after Tina's monologue, in which she sang a duet with also-pregnant Maya Rudolph to their unborn children. While they sang, I chugged a Red Bull. (I still want that endorsement deal.)

I swung by the cue card station one more time to read the latest incarnation of the script. I'd suggested a minor change to my lines, and I was really happy that Seth liked it and used it.

I walked under the bleachers past cables and wires and staffers and monitors and equipment—it was like walking through the actual guts of the show—and parked myself by the main entrance to the Hall to wait for the cue. Jim Carrey, who'd hosted a few weeks earlier, had dropped by to watch the show. Jim gave me a hug, and we chatted for a minute about the presidential reunion piece that Ron Howard directed for Funny or Die, Will Ferrell and Adam McKay's video Web site, the year before, Jim doing a fantastic Ronald Reagan, Will Ferrell as W., Dan Aykroyd as Nixon, Dana Carvey as George H. W. Bush, Chevy Chase as Carter, Fred Armisen as Obama, and me as, of course, Clinton.

Kenan was in one of the two change booths right by the stage, where you went when there wasn't enough time to go back to the quick-change booths down the hall. There was also a booth called "Pardo's booth" and another called "Paint Cans" past the cue cards;

a dresser might tell you to go to one or another between sketches based on which stage you would be due on next. Kenan was surrounded by a handful of hair and makeup people attaching the elaborate facial and scalp pieces that made up his costume as Jimmy McMillan. In the sketch, McMillan, in keeping with all the birther controversy that had been trailing Obama since his election, confirms rumors that he was born a billy goat. I don't even know who he is, but the audience loved it.

In another host booth, the curtains were drawn, so all I could see was a series of hands and elbows emerging out the top and sides. When Gena called for the debate cast, Tina emerged in full Palin makeup and hair, trailed by four or five women who'd been crammed in there with her to get her ready. They'd put her in impressively high heels, which meant, at six months pregnant, she had to be escorted on and off the stage so she didn't fall.

For this third and final version of the day, my gold-columned podium had acquired the word TRUMP in big letters across the top, mimicking the way his buildings were designed. When it came time for me to say, "Except you, Sarah Palin. I like you," and I paused the way we'd discussed a half hour earlier, the studio audience took the bait and applauded. In fact, I got two applause breaks that night. Sweet.

When I came offstage, Seth chased me down the hall. "That was terrific. It's great to have you back," he said, shaking my hand.

The truth is, I was propped up, as I'd always been, by an incredible staff of Emmy winners—hair, make-up, costumes, writers, producers. I don't care how great a person is, you put him out there without great material, he won't look good.

After I got back into my street clothes, I swung by Marci's office to say thank you. She was having a glass of wine with Renée Zellweger, who gave me a kiss. Marci said, "I love you. Stay for the good-byes tonight."

There was at least half an hour of the show to go, so I hung out in the hallway, chatting with people. While Ellie Goulding, the musical guest, sang her second number, I saw four African American men dressed in gray wigs and ladies' housecoats walk away from the studio. A moment later, I understood when I heard Gena say over the PA system, "Tyler Perry sketch can-celled." That was always the risk of being in a sketch slated for the later part of the show—you might get cut if they ran out of time.

Except for my last night as a cast member and the night McCain hosted, I hadn't stood on that stage at the end of the show in years. But I was feeling good. I knew the Trump bit had been a hit. So I not only got onstage, I stood in front. And the whole cast was lovely, giving me hugs. It was like going home, but in the nicest possible way.

A few days later, the real Trump announced that he wouldn't be running for president after all. And that's fine by me. But Steve was right—the photo of me and Tina was the one that ran.

Tru opened at the Bay Street Theatre on June 4. Judith Ivey had left the production a few days earlier, so it was down to me and Matt McGrath to make the final preparations. The first reviews were pretty good, and I was grateful. It was even more rewarding when Capote's old friends, like the actress Anne Jackson, came backstage in tears, telling me they felt they'd seen the man reincarnated in front of them.

And yet I made the mistake of watching the first half of Philip Seymour Hoffman's stupendous Oscar-winning turn as Capote in the eponymous 2005 film, and was filled with enormous doubt about how I was portraying him. I would note some of the gestures Hoffman used, and kick myself. *Why didn't I think of that?* I had to remind myself that we were, in fact, playing two very different versions of the man. Hoffman's Capote was young, vital, and at the top of his intellectual and productive abilities, about to produce

In Cold Blood, the masterpiece that would define his career. The Capote I was playing was near the end of his life, secluded in his United Nations Plaza apartment on Christmas in 1975, ostracized by the society dames who felt he'd betrayed them when he published a thinly veiled story about them in *Esquire*. Even his longtime partner, Jack Dunphy, had decided to spend the holidays in Europe without him. In his misery, Capote spends the duration of the play pickling himself to the point that he can barely speak.

During the show on opening night, I heard a cell phone ring in the front row. In character, I turned to the offender and said, "You on a waiting list for a kidney?" I was mortified when I realized it was Joy Behar, whom I admire a great deal. Sorry, Joy.

At the end of the play, I received a standing ovation. It was fantastic.

And yet some things never change. Can we talk about the sadistic fuck who invented Wednesday matinees? The Hamptons is a happening place on the eastern end of Long Island where celebrities and the well-to-do—think Alec Baldwin, Steven Spielberg, Martha Stewart—mingle during the summer months, going to lovely restaurants and clubs and the theater— *at night.* The daytime performances are for the elderly and infirm, people whose physicians have advised them against exposure to sun, wind, sand, heat, air, waves, and small children flying kites. These snow-capped gnomes hobbled in on canes and walkers, or arm in arm like an osteoporotic chain-link fence trying to remain upright against a gale-force wind, settled themselves in the first rows of the theater, and promptly fell asleep. The only thing that snores louder than an old man is an old woman. Some of the elderly simply sighed audibly or whimpered like newborn puppies. It was just like being back on those cruise ships again.

A couple of weeks into the show's run, I was half ready to quit if I couldn't persuade the theater to cancel the matinees. "You don't understand, I had a guy *die* in the middle of one of my performances under similar circumstances. I can *not* be thinking about whether members of the audience are expiring before my eyes while I'm trying to get through ninety-eight minutes of monologue."

The theater is very small, and the set was designed to make the audience feel like it was sitting in Capote's living room with him. It was a very cool layout, but during one of these horrendous matinee performances, an old woman tottered across the stage heading for the bathroom just as I was about to launch into a speech in which Capote contemplates killing himself. I tried not to let her distract me, but when she came shuffling back across the stage a few minutes later, it took a monumental act of fortitude to keep my composure. When the show was over, the audience woke in unison, struggled to their creaky feet, and gave me a standing ovation. They didn't even see the play.

A few weeks later, I was at Yankee Stadium with my friend Julie. She texted Michael Kay, the Yankees announcer, that I was there. Next thing I know, I'm looking at the Jumbotron—Al Roker, then David Gregory from *Meet the Press*, and then my face is one hundred feet tall overlooking the stadium. How cool is that?

I do have to keep going to meetings, and I do have to keep examining my thoughts, but I'm happy a lot of the time. And no more drinking, hopefully.

But free? I don't know. I'll be on my deathbed— many, many, many years from now—trying to make out the faces of my nearest and dearest through the fog of my imminent demise, and I'll say, "Well, I guess this is it. I'll see you on down the road." And the nurse will lean in and say, "Now say it like Bill Clinton."

Acknowledgments

Many people have helped me in ways both small and large with my career, my life, and this book. I am grateful to each and every one of them. No doubt the list that follows is incomplete.

At *Saturday Night Live*, thank you to Lorne Michaels for rerouting my destiny; Marci Klein for discovering me; Tina Fey for making me look better at sketch comedy than I really am; Mary Ellen Matthews for the awesome cover photo; Dana Edelson for photo research; Kenny Aymong; Steve Higgins; Mike Shoemaker; Katreese Barnes; Louis Zakarian; Jodi Mancuso; Lindsay Shookus; Bettie Rogers; Will Ferrell; Tracy Morgan; Gena Rositano; and both Jimmy Fallon and Maya Rudolph for holding my baby so much.

I am also grateful to my manager, Tim Sarkes at Brillstein Entertainment Partners, and to my agents at William Morris Endeavor, Stacy Mark and Mel Berger, with an assist by Graham Jaenicke.

At HarperCollins, gratitude to Jonathan Burnham, David Hirshey, Barry Harbaugh, Jaime Wolfe, Rachel Elinsky, Archie Ferguson, and Miranda Ottewell.

I would also like to thank:

Elizabeth Stein for putting up with me for a year during the writing of this book. Maris Kreizman and Joan Kasarda for their terrific transcription work;

Estee Adoram at the Comedy Cellar; and Caroline Hirsch and Linda Smith at Caroline's; Alan Spector for making my radio career possible; and the Westies for encouraging me to do stand-up in the first place;

Karen Giordano for making my Clinton more than an impression; my assistant, Lisa Robicheaux; my trainer, Matthew Grace; Wayne Tyson; Kathryn Tyson; Janet Collester; Ken Shirek; Craig Eden; Larry Laskowski; Frank Facciobene; Ginny Ballard; and Amber Paul, the yoga teacher who coached me out of my final darkness;

The doctors who have saved me on more than one occasion: Anne Griffin, Sally Burton, Andrew Ramsey, Fredda Gordon, and Nabil Kotbi. Along the same lines, I am grateful to the staff of the Emergency

Room at New York Hospital and the nurses of the NBC infirmary;

Tony Robbins, Martin Seligman, David Burns, and Ned Brown, all of whose books helped guide me during the tough times;

My inspirations: Martin Luther King, Jr., Harriet Tubman, Derek Jeter and Mariano Rivera for showing me how the game should be played, and Willie Mays for the catch;

The NYPD and the New York District Attorney's office for saving my family;

Eddie Galanek for convincing my dad that God loved him;

Donald Trump; Chris Matthews; Laura Ingraham; Maria Bartiromo; Senator John McCain; and Brucy Bochy. President George W. Bush for the invite to play catch. President Bill Clinton; Vice President Al Gore; and Vice President Dick Cheney for reaching out to me and showing me so much respect. Lynne Cheney for her hospitality. Whoopi Goldberg for understanding me. Howard Stern for telling people I was funny and giving me a chance. Tom Shales of the *Washington Post*. Whoever put me on the sand pile at the Ravenite Social Club. The woman who helped me get on the plane back to New York. The mayor at the Sanctuary, for hope and inspiration. Stanley the Bahamian guard. Hitchcock the Crow;

Truman Capote and Murphy Davis; and Joy Behar for the cell-phone line as well as her overall support;

And my daughter for introducing me to the miracle of unconditional love.

P.S. This book was written at the Gracie Mews Restaurant. Try the tilapia.